T0243452

Turning the Giant is a masterful blend of personal and professional insights. John Berra, one of the most respected leaders in industrial automation, has turned decades of experience into an eye-opening approach to business that should be a mandatory read for today's corporate executive. *Turning the Giant* thoughtfully uncovers the most important leadership traits to have today—and for generations to come.

STEPHANIE NEIL

Editor-in-Chief, OEM Magazine;

Former Senior Editor, Automation World magazine

John and I worked together at Emerson for over twenty-five years. A truly unique leader whose enormous passion and commitment to his customers were infectious, John brought to fruition transformational innovations and unwavering focus to make things better—both for Emerson and the automation industry at large. As Emerson's Chairman and CEO, I witnessed John's drive to never give up on his vision to make his customers more competitive and successful. If I said no to an investment request, he would keep coming back until he succeeded in convincing me. *Turning the Giant* is full of lessons that I wish all my leaders understood; to John, it was his life, and he lived these lessons thoroughly and authentically. In so doing, he made a huge difference to all he touched.

DAVID FARR

Retired Chairman and CEO; Emerson Electric

John Berra's wisdom and business acumen provide a serious lesson for leaders who strive for excellence in a large, complex organization. His heartfelt advice and practical guidance make this a "must read" for anyone who wants to succeed and make a difference.

ALAN BOECKMANN

Former Chairman and Chief Executive Officer, Fluor Corporation

I have had the pleasure of knowing John Berra for a couple of decades. I have always admired John for leading disruption in our very conservative industry. He is a great communicator and was always easily approachable. *Turning the Giant: Disrupting Your Industry with Persistent Innovation* is an excellent and timely read in today's challenging business environment driven by artificial intelligence (AI) and sustainability. I believe anyone seeking success in the business world and with an interest in making this world a better place will enjoy reading it.

ANDY CHATHA

President, Automation Research Corporation

Hundreds of books and articles have been written about change management. But many fewer resources provide real insight into change strategy. John Berra's experiences and analysis, conveyed in a delightfully conversational tone, will provide any business leader with suggested strategies as to how to effect change within their organization. And while fixing institutions that perceive themselves as broken is rarely met with resistance, changing the course of a large-scale enterprise that has become complacent requires both clarity of vision and a deft hand. *Turning the Giant* provides guidance as to how to achieve both.

AARON F. BOBICK

Dean, James McKelvey School of Engineering;
Washington University in St. Louis

TURNING THE GIANT

DISRUPTING YOUR INDUSTRY
WITH *PERSISTENT INNOVATION*

JOHN BERRA

Forbes | Books

Published by Forbes Books, Charleston, South Carolina.
An imprint of Advantage Media Group.

Forbes Books is a registered trademark, and the Forbes Books colophon is a trademark of Forbes Media, LLC.

Printed in the United States of America.

10 9 8 7 6 5 4 3 2 1

ISBN: 979-8-88750-242-7 (Paperback)
ISBN: 979-8-88750-221-2 (Hardcover)
ISBN: 979-8-88750-222-9 (eBook)

Library of Congress Control Number: 2023919681

Cover design by Matthew Morse.
Layout design by Wesley Strickland.

Since 1917, Forbes has remained steadfast in its mission to serve as the defining voice of entrepreneurial capitalism. Forbes Books, launched in 2016 through a partnership with Advantage Media, furthers that aim by helping business and thought leaders bring their stories, passion, and knowledge to the forefront in custom books. Opinions expressed by Forbes Books authors are their own. To be considered for publication, please visit **books.Forbes.com**.

To my wife, Charlotte, who has been my unwavering supporter.
The journey from college to retirement would not
have been possible without her.

Perseverance is usually described as a great virtue, but it has two aspects. Perseverance with an eye on the future, perseverance towards a definite objective is a great virtue: perseverance with an eye on the past is an equally serious vice.

WINSTON CHURCHILL

The real voyage of discovery consists not in seeking new landscapes, but in having new eyes.

MARCEL PROUST, LA PRISONNIÈRE

CONTENTS

FOREWORD

By Kathy Button Bell, retired Senior Vice President of Emerson, Chief
Marketing Officer and member of the Office of the Chief Executive

I am not an engineer. I don't even play one on TV. However,
as the head of marketing at Emerson for nearly twenty-five years, I have
been the official writer, painter, brander, musician, and sometime psychologist for one of the greatest engineering companies on Earth. John
Berra's vision was one of my first marketing projects, and he became the
professional partner behind my personal success as well as Emerson's
over the past few decades. He is the father of delivering holistic automation to critical industries. He introduced "new to the business"
and "new to the world" technology and business concepts that transformed automation in the energy, chemical, power, and pharmaceutical industries.

Turning the Giant is an ode to divining, introducing, and surviving
disruptive technology in a large corporate environment. John exem-
plified technical imagination, organizational creativity, vision, and
intense focus in a bureaucratic industrial company while serving
historically conservative industries. The book is his personal recol-

lection of approaching skeptical ears with eight innovative concepts, capturing in detail the emotional travails and the greater lessons to the next disrupter in how to succeed implementing brand-new ideas against the odds.

John had a clear and compelling vision of how to integrate Emerson's large portfolio of automation businesses to better serve the customer's pain points. He helped me persevere and survive integrating our autonomous businesses collected through acquisitions. John and I became fast partners with the identical vision of a much more powerful, unified company. I wanted to simplify the brand architecture to make Emerson easier to understand and do business with. He needed to integrate our systems, software, intelligent devices, and valves to help customers optimize their production while greatly lowering costs. John deserves much of the credit for moving us from being a pure component manufacturer to being a provider of complete automation solutions.

While John was a heralded technical leader across our served industries, many might miss the different organizational teams and structures he created to deliver on his vision. For disruptive software innovation, he created the highly independent technology "Hawk Site." Here in Austin, fifty hand-chosen, high-potential developers and customer-driven marketers were isolated from the bureaucracies of daily business to create DeltaV. Then he faced the internal skeptics and hired fifty killer-bee sales people to sell his new PlantWeb ecosystem. Finally, he kept the most consistent senior management team the Process Executive Group (or PEG) had had over nearly a decade, for consistent delivery of a single strategy to differentiate and change the process industries forever—making Emerson #1 in almost every category of *Control Magazine*'s annual Readers' Choice surveys.

Perhaps John's most enduring lessons contained herein are regarding the need to participate, engage, donate to, and manage the market. He emphasizes the importance of actively shaping and setting industry standards. He always believed in sharing technology to create open standards that made our industry better for all, and he highlights the benefits of participating in broader discussions, panels, and media events. He discusses the importance of utilizing trade, global business, and social media to share new and disruptive approaches, which offers not only increased reach but also improved credibility and opportunities for productive debate.

Finally, John gives the reader a personal prescription for leadership during disruption. He outlines key traits a leader must cultivate, from vision and passion to influence and action. John taught me the importance of tackling one skeptic at a time, slowly turning opinion toward the goal while listening and adding good new input to both the story and the offering. He was and is a champion for continuous improvement, and he understood from the beginning that the number one thing our competitors feared was if Emerson leveraged and unified all the strengths the different businesses had to offer. Today, his vision is nearly complete. Emerson has become a singularly focused software and solutions automation powerhouse. His mother, Geraldine, would be so incredibly proud.

INTRODUCTION

Everyone has giants. And these giants come in all shapes and sizes. Some giants overwhelm us, while others are manageable. There are giants in our personal lives, giants in our relationships, and giants in our careers.

In many self-help books, the common wisdom is to *slay your giants*. It's the whole David and Goliath analogy that life only gets better *after* you stand over your giant with a sword. But many times the giants that stand in our paths are obstacles to *turn* and not adversaries to *kill*. While the idea of eliminating a giant is appealing, it's not one grounded in reality. Most giants won't disappear with a David-like slingshot and a sword. Instead, we must learn to leverage them for our benefit so that they work in our favor.

When I reflect on my forty-plus years in the automation industry, I look at each phase of my life as a giant I had to turn.

My giant-turning days started in 1969 when I received a BS in Systems Science and Engineering from Washington University in St. Louis and began my career as a control engineer at Monsanto. And they continued in 1976 when I joined a measurement company named Rosemount, where I held several management positions, including president of the Industrial Division. And when I was named president of Fisher-Rosemount Systems in 1993 and promoted to senior vice president and process group business leader for Emerson Electric in 1999, a whole host of new giants emerged.

The greater my position and influence, the greater my giants. And during my first few years out of university, I'll admit I was eager to slay my share of them. I had a relentless work ethic and never took no for an answer. But over time, I learned to channel my energy in the right way and turn the giants I faced to my advantage. In the words of leadership author Andy Stanley, I realized my giants weren't "problems to be solved" but "tensions to be managed."

I'd always have challenges. But I had to learn to turn them to my advantage. I say this because if you just looked at the highlight reel of my life, it's easy to think I'm a guy who didn't have much adversity to overcome and that my meteoric rise was the result of fortunate events and timing. But the truth is, I've had to turn giants in each company I've worked for in the automation industry.

This meant I drove the development of two important automation industry standards. The first one was the hybrid addressable remote transmitter (HART) protocol. And the second involved being an early proponent of an all-digital fieldbus, serving as the chairman of the board of the Fieldbus Foundation from its creation in 1994 to my retirement in 2010. From 1988 to 1990, I was chairman of the board of the Measurement, Control, and Automation Association. In addition, I have served on the board of directors of Ryder System, Inc. (NYSE:R) and National Instruments (NASDAQ:NATI).

While these titles might not mean much to someone outside the automation industry, inside these circles they represent a great deal. And because of my efforts, a few awards I've received include the following:

- The Emerson Electric Technology Leadership Award

- The Washington University Alumni Achievement and Distinguished Alumni Awards

- The Lifetime Achievement Award from ISA

- The Frost and Sullivan Lifetime Achievement Award.

In 2003, I was named one of the fifty most influential industry innovators by *InTech* magazine, and in 2011, I was voted into the Process Automation Hall of Fame.

By most objective standards, my life has been a "success." But it certainly didn't start off this way. While it's tempting to think I grew up on third base, the truth is I came from humble beginnings. Both sets of my grandparents came to America through Ellis Island. They sought a new life, and they found one. They knew what it was like to experience prosperity and understood what it felt like to stumble into the Great Depression and struggle to make ends meet.

Despite these setbacks, they kept pressing forward, and it's not hard to discover how I inherited the giant-turning gene. My father, Joe Berra, was a World War II veteran who fought in some of the worst battles in the Philippines. While he never talked about these experiences, I now realize the toll it took. When he returned to America, like many veterans, the only thing he wanted was a nice home with a white picket fence.

And he found this lifestyle with my mom, Geraldine. After meeting when dad was home on leave during training, they corresponded by mail. Despite scarcely knowing each other, they courted one another and wrote a combined 109 letters—each one I have stored in my office today.

Sometimes I read through these notes and marvel at their perspective. After all my father experienced, having a job and a quiet life meant everything to him and my mom. And throughout my life, when my career took me from one position and city to the next, there were times they thought I was crazy for pulling up roots and moving somewhere new. Despite having different views on how to handle career advancement, their commitment to certain values and hard work left an indelible imprint on my life. Without their sacrifice and my wife Charlotte's support, I wouldn't have turned half of the giants mentioned in these pages.

Today, I'm retired. Some things are different, and some things remain the same. I still get seventy-five emails a day, but instead of questions from inquisitive employees, I receive advertisements for walk-in tubs and Medicare Advantage plans. I still get together with my colleagues, but instead of meeting at the office, we hang out at the golf course or on a fishing trip.

I've been out of the game for a few years, and sometimes my head spins when I see the speed at which technological advancements are made. I'm also well aware that my understanding of advancements in artificial intelligence today might prove obsolete in five minutes. But when I think back to my career in the automation community and those technological innovations made during the '70s, '80s, '90s, and early 2000s, I realize the principles that helped me turn those giants are the same principles that can work for you today.

While each generation looks very different from its predecessor, there are many similarities. Each generation has its own set of giants. But the way we turn them often requires a similar approach. And this is where I believe I can help.

Turning giants is wonderful.

In my business, there was nothing better than turning a strong "no" into a "yes." To this point, I remember this time when I was about a decade into my career. I was working in marketing at a company called Rosemount when I received word that a potential client of ours, a gas plant in Liberal, Kansas (which is situated in the middle of nowhere), had made the decision to go with one of our competitors. To me, this didn't make sense. I knew we were the best in the industry, and so I phoned the manager and asked if I could fly over in person to meet with him.

He said yes, but that I'd better get there soon. The next day, I was on a plane and arrived at his facility. As we spoke, I probed him for information and asked why he thought it would be best to go with our competitor.

After a bit of hemming and hawing, he got down to the real reason—recognition. He was new to his job and wanted to make a major splash. By going with an established company like Rosemount, he'd be making the move everyone expected. And this meant he wouldn't receive the glory and positive newsletter write-ups he desired.

It was then I hit him with a question he didn't expect. "Have you thought about what kind of write-up you'll receive if your decision *doesn't* work out?" I asked.

He paused.

"Rosemount is the number one supplier of measurements in the world," I continued. "So why would you risk your career on this decision?"

I could tell his wheels were turning. And after a few more minutes of conversation, something clicked and he said, "OK." And right then, he walked out to his secretary and told them to cancel the order they'd made for this other company and to go with Rosemount instead.

I always relished these kinds of situations.

Some people look at the whole idea of "turning giants" and think to themselves, *that doesn't sound like much fun*. And to be clear, turning giants does require a lot of effort. But it's also one of the greatest thrills you'll ever experience.

I write this book not as someone who is a celebrity. In fact, if you're not part of the automation community, it's doubtful you even know who I am. Like you, I've read my share of business books by famous individuals. And while many of them were helpful, they were often either too academic or detached from how the real world operates.

In *Turning the Giant*, I've adopted a different approach. I've resisted the urge to include a bunch of graphs and charts. In fact, if you're from the automation community, it's likely you'll think I could have said a lot more on certain topics. But my goal is to make these principles accessible to everyone.

Throughout my career, I've experienced so many wonderful, real-world lessons. And now I want to share from my successes and mistakes. By doing so, I want to help you, the reader, become a more effective giant-turner in your organization.

So, turn the page, and I'll show you how to start turning your giants.

CHAPTER 1:

THE GIANT OF CORPORATE BUREAUCRACY

Have you ever worked at a company and felt frustrated?
Each day feels like a carbon copy of the one before. Management always asks you to turn in reports that seem focused on the wrong ideals. Any idea you present slowly drowns in a sea of red tape. The next deadline feels like it will make or break you. And the culture in which you're supposed to be a high achiever is less than ideal.

It wasn't always this way. You started with high hopes and thought you could be a catalyst for change. But maybe you're twenty years into your career, and it hasn't happened. Or you're just coming out of university with aspirations to climb the corporate ladder. Instead, you feel mired in a pit of corporate bureaucracy.

Internally, you're thinking to yourself: *Let's go! What are we waiting for? Why doesn't management seem to notice the obvious?*

If this is your story, you're not alone. In 2022, Gallup released its State of the Global Workplace report that included some staggering figures. It revealed that 60 percent of workers were emotionally detached at work and 50 percent of workers said they felt stressed at their jobs on a daily basis.[1] Certainly, some of this can be tied to the COVID pandemic, but there is another factor at play.

According to BBC's Ali Francis, "For decades, the cultural mandate in many Western countries has been hustle hard for your employer, and you'll be rewarded."[2] Work hard, keep your mouth shut, do what you are told, and a hefty paycheck at the end of the rainbow awaits. But as Francis notes, younger workers want something different. She writes:

> Having observed older workers experience burnout, time poverty and economic insecurity at the grindstone, they're demanding more from workplaces: bigger pay cheques, more time off, the flexibility to work remotely and greater social and environmental responsibility. Many of these values were millennial preferences, but for Gen Zers, they've become expectations—and they're willing to walk away from employers if their needs aren't met.[3]

No one likes bureaucracy, and no leader thinks of themselves as bureaucratic. But like it or not, the giant of corporate bureaucracy looms large in most established organizations. And often, we do not have the luxury or desire to pack up our stuff and start a new job. So, when our backs are against the wall, *what do we do? Where do we turn? And how do we succeed?* These are the questions I've spent a lifetime attempting to answer.

---

1   "State of the Global Workplace," Gallup, accessed May 10, 2023, https://www.gallup.com/workplace/349484/state-of-the-global-workplace.aspx.

2   Ali Francis, "Gen Z: The Workers Who Want It All," BBC, accessed May 10, 2023, https://www.bbc.com/worklife/article/20220613-gen-z-the-workers-who-want-it-all.

3   Ibid.

WANTING SOMETHING DIFFERENT

I've always possessed a strong drive to succeed. Part of this goes back to my childhood. I didn't grow up in a home with a lot of money. After Dad got back from fighting in the Philippines, he and Mom got married and brought me into the world around two years after his return. Because their entire courtship was conducted through letters, my parents barely knew each other when they said, "I do."

For the years leading up to their marriage, they lived two very different lives. And there was no way a few dozen letters could explain some of the traumatic horrors of war he witnessed. As a result, when Dad returned to America, he was a very different man than when he'd left. Today, we would say Dad struggled with post-traumatic stress, and it was tough for him to adjust to civilian life. And we never had much money.

Looking back, I have so much sympathy for what he experienced. But as a young teenage kid, all I knew was I wanted to become successful so that I could provide a better life for my family. You could say I lived with a bit of a chip on my shoulder. I was ambitious and determined to make something of myself.

I received a national merit scholarship to Washington University in St. Louis and started my first semester in the fall of 1965. While other kids were content to live with their parents and attend school during the day, I did everything I could to branch out and live on campus for my junior and senior years. I wanted to maximize my educational experience. Tight on funds, I was happy to find out the treasurer and president of my fraternity group received free room and board. And so I ran for and was elected to treasurer in my junior year and became president in my senior year.

My time at Washington University was exhilarating. At first, I wanted to be a chemical engineer, but organic chemistry changed my mind. System science became my major, and I loved it. At its core, system science is all about automatic control. And a great example of system science is self-driving cars.

Many of the professors were legends in their field. And to broaden my horizons, I even took a few additional courses outside of engineering. My first taste of leadership came when I was elected president of my fraternity pledge class. And during my four years, I was also active in campus activities and chosen to be a member of both the junior and senior leadership honoraries. But the most important part of my time at Washington University was when I met Charlotte, the love of my life. As of this writing, we have been married fifty-four years.

ENTERING THE WORLD OF BUREAUCRACY

In the spring of 1969, I graduated with a BS in Systems Science and Engineering. This marked my transition from academics and into the world of corporate bureaucracy.

By the end of my last semester, I was excited, ambitious, and eager to change the world. And so, I lined up a series of interviews with companies I respected. The first was with Pratt & Whitney. Today, their commercial aircraft engines "power nearly 30 percent of the world's mainline passenger aircraft fleet."[4] And at that time, they were a giant in their industry. But unfortunately, they were also a giant of corporate bureaucracy.

I'm sure much has changed since I first visited one of their facilities. But in 1969, everything was systematized and controlled from

---

4   "Products and Services," Pratt & Whitney, accessed August 4, 2023, https://prattwhitney.com/products-and-services/products.

the top down. I was interviewed that day with several others, and as I toured the facility, I grew more uneasy. I looked at the engineering pool where I was slated to work, and to me it felt more like a factory. There were no cubicles or partitions. Each engineer sat in this large open room with his desk butted up tight against another and arranged in long rows. One chair, one desk, and one telephone. And that was it.

The thought of working there each day horrified me. Engineers were nothing more than cogs in a wheel, and they looked to me as expendable as the desks at which they worked. When engineers spoke on the phone, eight other people could listen in on their conversations. And to make matters worse, there was an open second level above, making it possible for managers to make sure everyone was doing what they were supposed to do. There was no room to think, let alone be an individual.

But such was the culture of that time. In subsequent interviews, I encountered this same basic structure.

THERE HAS TO BE A BETTER WAY

Two weeks after graduation and several interviews later, on June 5, 1969, I reported for my first day of work as an instrument and control engineer at a company named Monsanto. At the time, Monsanto had a large corporate engineering department with eight hundred engineers from different disciplines.

Unlike the other companies I'd visited, each engineer at Monsanto had a cube with a reasonable amount of workspace. The people were great, and the company emphasized training and development— assigning me a mentor who taught me all I needed to know. Aside from these benefits, I was happy to be in the chemical industry and not in aerospace. Just over a month after I started at Monsanto, Neil

Armstrong set foot on the moon. And as I predicted, after this historical accomplishment, the aerospace industry went into a period of national decline while the chemical industry flourished.

Monsanto was a classic American company that had been around since 1901. This lengthy history was part of the country's corporate DNA. Monsanto was a leading global supplier of herbicides and seeds. It also diversified into petroleum, fibers, building materials, and packaging. You might know them for their most famous product, *Roundup*. Monsanto was also a bulk chemical producer of sulfuric acid, phosphate, and fertilizer. As if that wasn't enough, it also produced aspirin and the thermoplastics polymer that went into Barbie dolls. Needless to say, Monsanto was a large and incredibly diverse company.

But with large comes that dreaded word, *bureaucracy*.

Sometimes I smile when I hear someone complain about their mundane job they cannot stand. I smile not out of condescension but sympathy. At Monsanto, I worked with a mixed group of engineers. Some were senior engineers who had been around for a while, and others were right out of school. There was an overall head of central engineering, but the way Monsanto was structured, engineers seldom spoke to anyone above their immediate supervisor. There was a clear chain of command.

And at the time, I had a supervisor who was interesting. He had a crew cut and wore these orange wing tipped shoes with leather heels and soles. Every time he walked, we'd hear this unmistakable clip-clop, clip-clop on the linoleum floor. And when we did, we knew the boss was on the way, and it probably wasn't going to be good!

This boss oversaw his engineering division like a commanding officer in the army. He had very little tolerance for any engineer whose personality landed outside the "bandwidth" of the personality he expected. Let's just say I was a couple of standard deviations outside his target of expectations.

As a new instrument and control engineer, I was introduced to the world of process automation. Visualize a NASA-like control room full of screens and operators in a manufacturing plant that had thousands of sensors monitoring factors such as flow rate, temperature, pressure, and composition. Everything being monitored was in tanks, vessels, and pipes. This meant you couldn't see what was going on and had to rely on the sensors for feedback.

Each sensor had a dedicated pair of wires running from the factory floor back to the control room.

CONTROL LOOP

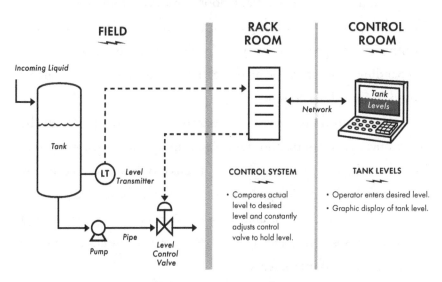

The control system compared the value the sensor was sending back to what the value was supposed to be. From there, the control system sent out a signal over a different pair of wires to a control valve that adjusted its position to bring the parameter to the desired value. It was a single loop with automated technology that monitored the parameters and made any necessary adjustments. And in the end, the fundamental purpose of automation was to control parameters in the plant to make the best possible product.

Then, on top of that, there were more advanced oversights adjusting on the fly for changes such as the weather or humidity. And if that wasn't complicated enough, the equipment in each control room at each big industrial company was entirely proprietary to the supplier. This meant each supplier developed their own unique method of monitoring and responding to what was happening on the plant floor.

Back in those days, even the operator stations were custom-built by the supplier. We called control systems "Big Irons" because they were so large and expensive. The system controlled the stuff that was manufactured on the plant floor. And this "stuff" could range from beer, to chemicals, to gases, or, in the case of pharmaceutical companies, drugs where every molecule was measured and traced. Depending on what was being made, the control-system requirements could be incredibly complex.

As a control engineer at Monsanto, my job was to design the instruments and controls for new plants and plant extensions. If that doesn't make sense to you, picture a thousand devices connected by a thousand pairs of wires going through junction boxes and terminal strips. Now, imagine you are the one responsible to ensure *each* of these wires was connected to the right terminals on either end and that there was no interference.

Here's what this would look like. With the help of a coworker, I would test each of these lines to ensure there was an accurate signal. On one end, I held a tone generator while my partner held a tone receiver on the other. And over a radio, we'd communicate back and forth. "John, I'm not getting a signal on this one. Can you try it again?" or "Yup, I got it. Let's move on to the next one." We called this mind-numbing process "ringing out the wires."

Day after day, I showed up and repeated this same process. It's hard for me to think of a more repetitive, less inspiring task. But the one thing this job gave me was time to think. Internally, I kept saying to myself, *there has got to be a better way!* However, I soon discovered innovation in large industrial companies is extremely difficult. The existing internal structure was a formidable obstacle, and management was seldom eager for change.

My three years at Monsanto left me feeling discouraged and unfulfilled. The company provided wonderful training, but they did not capitalize on this investment and continued to give fully trained engineers like myself the most menial tasks. This didn't sit well with me because I wanted more challenge and responsibility. But little did I know this experience at Monsanto would end up planting the seeds I needed to sprout a fruitful career.

EVERYONE NEEDS FULFILLING WORK

I'm guessing you have your own tale of frustration. Internally, you keep repeating this same line, *there has got to be a better way!*

There are a few reasons you feel this way. For starters, your frustration is an indication you care. You want to make a difference, but you feel like your hands are tied. And second, it's natural to want fulfilling work.

There is much talk these days about frustration and anxiety. In Johann Hari's book, *Lost Connections: Uncovering the Real Causes of Depression – and the Unexpected Solutions*, he reveals that one of the nine major reasons people struggle with depression is lack of meaningful work. Hari interviewed hundreds of people who had struggled with depression and anxiety, and one of the interviewees shared with him that "The worst stress for people isn't having to bear a lot of responsibility." Instead, it is having to endure, "work [that] is monoto-nous, boring, soul-destroying; [where] they die a little when they come to work each day, because their work touches no part of them that is them."[5]

We were made for meaningful work. Unfortunately, that is not how many organizations are set up to run. One man we have to thank for this is Frederick Taylor. According to the CEO of *Populace,* Todd Rose, "Frederick Taylor is probably the most important person that most people have never heard of."[6] Taylor, born in 1856, was the author of a book titled *The Principles of Scientific Management,* in which he wrote, "In the past the man has been first; in the future the system must be first."[7] Taylor coined the term "manager" and created a top-down structure where systems usurped the will of individuals. Thanks to Taylor, a system was created where we are all cogs in a corporate wheel, and it's the system that matters most.[8]

Throughout the twentieth century, many contributed to Taylor's vision. Henry Ford increased automation, and others developed the concept of continuous improvement for assembly lines. And now, the rise of robotics has already rendered many human jobs obsolete.

Personally, I knew I could never show up to work as a "crank turner." While some find satisfaction in showing up each day to complete a task they can do with ease, this was never the way I thought. I had to believe in what I was doing and have a way to contribute. In the words of Simon Sinek, I had to have a "why."

5   Johann Hari, *Lost Connections* (London: Bloomsbury Publishing, 2018), 83.

6   BigThink, "The Power of Vulnerability," YouTube video, 1:25, March 1, 2022, https://www.youtube.com/watch?v=E4erHIZpnyQ&ab_channel=BigThink.

7   Frederick Taylor, *The Principles of Scientific Management* (Mineola: Dover Publications, 1911).

8   BigThink, "The Power of Vulnerability," 2:00.

THE WORKFORCE HAS CHANGED

In many respects, my thinking in 1969 was well ahead of my time.

Today, whenever I speak to audiences who range from baby boomers to Gen Zers, I'm reminded of how different each group thinks. Whenever I mention terms such as "innovation," "meaningful work," and "challenging the system," I'll receive nods from young attendees. But when I speak about "staying late and showing up early" or "ways you can help your company succeed," I receive more affirmation from those who are older.

In my generation, workers asked questions like "What do I need to do to be a success?" "What do I need to learn?" "What do I need to focus on?" "How do I help the company and my career?" Today, when kids come out of school, they ask, "What can the company do for *me*?" "How can the company change to enable *me* to make a difference?"

Personally, I'm a mixture of these two generations. I believe in hard work and putting in the hours. I think it's important to prove your worth before demanding a raise. But I also understand our younger generation that wants to make an impact and "have a seat at the table." I admire this passion and energy. In truth, younger and older workers can learn from one another. Times have shifted, but certain principles remain the same.

Along with this shift in the workforce, there has also been a shift in management. With the rise of ESG (Environmental, Social, and Corporate Governance) and the #MeToo movement, the dynamics in a workplace have changed, and there is an added emphasis on diversity and inclusion. In the past, managers had EEO (Equal Employment Opportunity) reports they had to fill out. But today it's much more complicated. Government agencies want to see a company's sustainability, and it's become increasingly common for organizations, customers, and even governments to all have a say in how one runs their business.

I bring this up not to say some of these reforms haven't been helpful. But I do so just to point out the landscape has shifted. To be frank, management runs a lot more afraid today than it did in the past. They're afraid a simple mistake or misunderstanding could wind up on social media and cost them dearly. Consequently, while some leaders have become more inclusive and trusting, others have become much more careful with whom they allow in their inner circle.

I share this not to demotivate you but to raise awareness of how bureaucracy operates today. There are fresh challenges. But with these challenges comes great opportunity.

HOW TO NOT BE A BUREAUCRAT

Albert Einstein once said that "bureaucracy is the death of any achievement." If you're someone who has achieved some success and you now lead at a high level, this makes you increasingly susceptible to becoming the thing you always despised. While few people think of themselves as bureaucratic leaders, here are a few questions to consider as you guard against going down this perilous path:

- When was the last time a subordinate challenged your view on an important issue, and this caused you to change your mind?

- Do workers two or three levels below you have access to you?

- Have you developed relationships with a few people who will come to you privately and tell you when and how you have messed up?

- Do you hold yourself to the same standard you hold your employees?

The reality is that as the modern workforce grows more global, companies that have a bureaucratic culture will be left behind. For example, I live in Austin, Texas—a hotbed of the tech industry. If you're a worker who has managed technology, you can leave your organization at any time and find ten other companies to hire you. The grass is always greener somewhere else. And personally, I don't begrudge employees for leaving unhealthy organizations that do not value them as individuals.

As Tom Peters and Bob Waterman write, "[M]ost of the management systems that treat people as 'factors of production,' as cogs in an industrial machine, are inherently demotivating. People are wonderfully different and complex. Leaders need to set people free to help, not try to harness them."[9]

If you are a leader, you are responsible for setting the culture in your organization. And as I've discovered, one of the primary causes of turnover is internal politics. Many organizations have a culture of backstabbing and lack of cooperation. And as I like to say, the only person who wants to work in an organization like that is someone who *is* that kind of person.

With that said, there are three primary ways you can safeguard against being a bureaucrat.

FIRST, LISTEN TO THE CONCERNS OF THOSE ON YOUR TEAM

Create a culture where workers can share their honest concerns and not be punished. Fast-forward a couple of decades to when I became the new president of Fisher-Rosemount Systems. I was working in my office as my chief financial officer (CFO) held a meeting with a

---

9   Tom Peters and Robert H. Waterman, *In Search of Excellence* (New York: Harper & Row, 1982), Author's Note.

group of potential clients next door. As they spoke, one of the guests said, "Well, what would John think about this?"

My CFO responded, "I'm not sure. Let me ask him."

Every head in the room shot up. "You mean you can just walk into his office and ask him?" one of them replied, as though this were a foreign concept.

"Of course," my CFO responded. And two minutes later, he was back from my office with an answer.

I can't even remember our topic of conversation, but I'll never forget when my CFO told me about the shocked expressions. The idea that the president of a large organization could be approached without some bureaucratic process was something outsiders couldn't comprehend. But this open-door policy is one I've carried for years. I always wanted to hear what was going on in my organization—even if this broke the chain of command.

SECOND, ENCOURAGE AND REWARD
DISRUPTIVE INNOVATION

Around this same time at Fisher-Rosemount Systems, I recall having a meeting with a dozen or so department leaders. I asked each of them to provide an update. And as they spoke, nearly each talked about how they were coming in under budget. Finally, after the last person spoke, I said to the room, "I don't understand. Each of you are happy you're under budget. But this past quarter, we *lost* $10 million."

I went on to point out that budgets are only good for about three days. Instead, the real goal is to focus on profit and loss. And doing so would require some disruptive innovation. Business couldn't be conducted as usual. Something had to change. I wasn't interested in rewarding leaders for coming in below budget. I wanted them to be disrupters and advance the overall interests of the company.

THIRD, LEARN FROM OTHERS WHO
HAVE WALKED BEFORE YOU

When I first started at Rosemount, I was about as green as green can be and even struggled to read our company's profit and loss (P&L) statements and balance sheets. Not knowing what to do, I turned to the head of accounting, Gene, and asked him if he would be willing to meet for coffee and teach me the ropes. He graciously accepted and even gave me a few books to study. Two years later, I danced my way through any P&L statement and balance sheet in our organization. But this only happened because I wasn't too proud to ask for help.

STEPS TO TURNING THE GIANT OF CORPORATE BUREAUCRACY

Corporate bureaucracy is not an obstacle from which to run. It is a giant to turn. Every organization has some level of bureaucracy that involves a hierarchy and structure of approvals. Bureaucracy will always exist. Accept that. But if you'd like to work through the giant of corporate bureaucracy, here are a few suggestions.

FIRST, UNDERSTAND HOW BUREAUCRACY
WORKS IN YOUR ORGANIZATION

Don't expect to change the culture of an organization in one day. Be strategic and study how and why the current structures are in place. If you can make better changes, great. But if not, you've got to work your way through any maze to accomplish what you need to do.

As you can see from the image below, every organizational chart is drawn with boxes and vertical lines connecting these boxes up the chain to the top. Those who want to bring about change need to work across these vertical columns. To do this, they must have the ability to influence people who do not report directly to them. And any change in an organization comes from hard work and connections across these "silos."

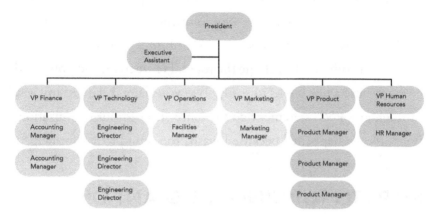

Here is how employees often get this wrong. Let's say Bob in accounting has an innovative idea that will impact not only his department but also those in operations. But rather than speaking to the head of operations, Bob takes his idea directly to the top. And that's when things start to get messy. Because Bob hasn't done his homework, his idea might get rejected simply because he hasn't considered certain variables and spoken with people who will be impacted when his innovative idea is implemented.

It's also important to understand the psychology of many managers. If you present a new idea, insecure managers will respond in one of two ways. First, if it's a good idea, they will want the credit. And second, if it's a bad idea, they will not want to take the blame. That is how many managers think. They see someone with a new idea as a threat. Recognize this, and adjust your approach accordingly.

SECOND, DEVELOP THE RIGHT ALLIES

Partner with others who share your beliefs, and agree with the changes you would like to make. Know the personalities of the people who are working alongside you. Remember that you want to *turn* people, not *change* people.

People might not want to go along with every idea you have, but you need to start somewhere. If you get a no, reevaluate your message. Know who you're speaking to and tailor your language for each conversation.

You don't always have to take your ideas to the top. And sometimes, after speaking to others, you'll realize you need a very different approach than what you first envisioned.

THIRD, COMMUNICATE YOUR MESSAGE EFFECTIVELY

Some people have the right idea, but they're horrible presenters. Their body language is poor, they struggle to speak in public, and they come across as incompetent.

But as many authors such as Daniel Goleman have noted, skills like emotional intelligence are often more important than intellectual intelligence. It's one thing to know what needs to change. But it's another thing to be the right messenger for that change. Fortunately, communication is a skill anyone can learn.

FOURTH, BE WILLING TO TAKE SOME RISKS

There is an illustration I've found helpful when interviewing employees. I'd ask them to take out a sheet of paper and draw a line down the center. At the top of the right side of the paper, I had them write, "Things I did well at my last job." And at the top of the left side, I asked them to write, "Things I didn't do well at my last job."

Sometimes employees were hard on themselves and listed more on the left than on the right. Others only talked about what they did well. But I always wanted to work with people who had a healthy list of things they did well and also had a few things they didn't. This showed me they weren't risk averse. They tried, even if it didn't end up working out.

FIFTH, DON'T RUSH TO THE OPPOSITE EXTREME

The opposite side of a bureaucratic culture is one where *anything* goes. I once had a head of operations work for me who wasn't used to answering questions. I'll call him Mike.

One day, Mike filed an appropriation request. After reviewing it, I had a few questions and sent him an email, asking him to provide me with some justification for his numbers and assumptions. To me, this was a standard operating procedure. I wasn't shooting down his idea. I just needed some clarity.

The next day, Mike walked into my office and closed the door. "John," he started, "don't you have any confidence in me?"

Taken aback, I asked why he felt that way.

"Well," he said, "I'm the head of operations. I ought to be able to judge what should and shouldn't be done. You should trust me to do my job."

"Mike, this has nothing to do with my confidence in you," I said. "I just need to know how you arrived at your decision because it's going to result in our company spending a lot of money. There is nothing personal about this." As I soon discovered, Mike's previous boss (my predecessor) was someone who rubber-stamped every decision. And that was one of the reasons I was asked to take over.

Bureaucracy is never the answer. But the opposite extreme can prove just as detrimental to the growth of an organization.

TURNING GIANTS IS FUN!

It's all well and good to talk about how you loathe bureaucracy, but at the end of the day, regardless of whether you're in management or attempting to "lead up," bureaucracy is something everyone must face. Bureaucracy is a giant. In the words of Clayton Christenson, "Incompetence, bureaucracy, arrogance, tired executive blood, poor planning, and short-term investment horizons obviously have played leading roles in toppling many companies."[10]

And like any dangerous giant, you do not make it go away through avoidance. Bureaucracy is something you must confront before you move forward.

Turning the giant of bureaucracy isn't easy. Some days it's downright frustrating. But as I've discovered, properly channeled frustration can become a spark for innovation. Anyone can complain, but it takes a person of character to overcome. To some, giants cause them to shrivel. But as Malcolm Gladwell pointed out in *David and Goliath*, "Giants are not what we think they are. The same qualities that appear to give them strength are often the sources of great weakness."[11] With every giant exists an opportunity to test our resolve and create a breakthrough others wouldn't have ever imagined.

It's this thrill of turning giants that causes people to sign up for tough events like Ironman competitions. If you ask me, running a full 26.2-mile marathon, biking 112 miles, and swimming 2.4 miles do not sound like my idea of a great time. But the reason people do this is because there is something exhilarating about testing the human spirit. The joy of facing a hard challenge and overcoming it is wonderful.

---

10   Clayton M. Christensen, *The Innovator's Dilemma: Management of Innovation and Change* (Boston: Harvard Business Review Press, 1997), 325.

11   Malcolm Gladwell, *David and Goliath: Underdogs, Misfits, and the Art of Battling Giants* (New York: Little, Brown and Company, 2013).

I understand giants can be frustrating. Some days, they don't feel like they're worth the battle. And in these moments, it's tempting to give up. But I want to challenge you to keep going. Run toward your giants and not away from them. However, to develop this kind of giant-turning mentality, it's imperative that we confront the greatest giant of all—the giant of self-doubt.

CHAPTER 2:

THE GIANT OF SELF-DOUBT

The only limit to our realization of tomorrow will be our doubts of today.

FRANKLIN D. ROOSEVELT[12]

While I've always tried to hide internal emotions, the truth is that for most of my life I have battled this giant of doubt. *Do I have what it takes to succeed? Am I the right person for the job? Can I handle the added pressure of a new position?* These are all questions I've asked myself a thousand times.

I've come to discover that maximizing my potential and facing self-doubt run hand in hand. I've always embraced a challenge. But this adventuresome spirit comes with a price. Because I wasn't content to learn a craft and practice it for several decades, this meant every few years I felt like I was in over my head.

Rewind to 1969. After leaving university and starting my job as an engineer at Monsanto, I was assigned a partitioned cube with a

---

12   Undelivered address for Jefferson Day, April 13, 1945, final lines, in Public Papers (1950) vol. 13, p. 616.

desk, some shelves, and a filing cabinet. It wasn't elaborate, but I felt a sense of pride that I had now entered the workforce. My first day on the job I was assigned a mentor named Lee, who took me under his wing. And the longer he spoke, the more I realized I didn't know. Doubt started to creep in. *What was I doing here? Shouldn't I still be in school? How do I know I'm up for this challenge?*

Engineering in the real world was very different from life in university, and I felt like a deer caught in the headlights. But Lee was patient and took time to explain each new aspect in detail. This was one of the reasons I loved Monsanto. Their culture of mentorship and collaboration meant new guys like me could learn from the likes of experts like Lee. We weren't thrown to the wolves on our first day, and we knew exactly what leadership expected of us.

After spending several months with Lee, I was sent for more training to a two-week school taught by senior engineers.

Monsanto training class. John is third from right, back row

During these fourteen days, we went through a whole gamut of automation devices. We learned new methods for calibration, testing, and control. In other words, I entered engineering heaven!

My engineering brain felt like it was on dopamine overload, and I couldn't wait to put what I learned into practice. This advanced training, which was rare among organizations in that day, gave me the confidence I needed to tackle my fears. It taught me that one of the best ways to overcome internal doubts was to gain more information and keep moving forward. When I left, I felt ready to tackle the world.

But then, a new challenge. As part of Monsanto's ongoing training, I was sent to spend six months in one of their company's US manufacturing plants. The plant I'd been assigned to was in Long Beach, California. And on day one, I was named the temporary maintenance engineer. This title, however insignificant it was, placed a fifty-pound weight on my shoulders. Once again, I felt like I was back in over my head. I was the new guy, the outsider from corporate engineering. And I was now responsible for technicians who had been at this location for years.

Thankfully, within a few weeks, I settled in and gained a ton of invaluable experience. There was something about seeing how each part of an operation ran. And to this day, I still believe there is no substitute for working on a plant floor. There is only so much you can learn in a classroom or a corporate office. But when you're out among the workers, you grasp how valuable every person is to the operation of a facility.

An added benefit was that this environment forced me to grasp the art of troubleshooting. I learned how to deal with problems on the fly. And during this time, I even set foot into the world of innovation by developing a new approach to controlling an impact mill—thus increasing throughput and decreasing downtime. A small victory but one that made me thirsty for more.

CROSSING TO THE SUPPLIER SIDE

As with most companies, there were perks and problems. While Monsanto was great at training, they struggled to empower non-chemical engineers to climb the corporate ladder. After a while, I was ready for more challenging assignments. But they did not come, and I realized my career had plateaued.

I'd received some phenomenal training, and yet I was still handed menial responsibilities. In my mind, I was a race car driver relegated to postman duties. This experience taught me a valuable lesson. The greatest training environment only goes so far. It must lead to greater responsibility. If it does not, people feel undervalued and demotivated. And so it was that after three years, I contacted a headhunter and started searching for another job that would let me spread my wings.

In 1972, I joined J. F. Pritchard in Kansas City and was thrown into the deep end. Whereas Monsanto had their own engineering, procurement, and construction (EPC), Pritchard was an EPC *contractor*, and this meant end users hired them to deliver turnkey projects. EPC companies are a huge part of automation, and I was assigned a project for an addition to a major refinery on the West Coast.

I wanted more responsibility, and more responsibility I received. My major task at this refinery was to supervise installation and startup. This field experience taught me some valuable lessons. And living in the world of automation engineers gave me a unique perspective when I eventually became an automation supplier. Automation engineers have the widest span of knowledge in their profession. They need to understand the entire process—mechanical, electrical, digital, networks, and software. I often say automation is a noble profession that just doesn't get the recognition it deserves.

There was so much to learn at Pritchard. But it wasn't long before the same story started to unfold. I had a great boss, but as at Monsanto, instrument and control engineers were second class to the chemical engineers and project managers. As a result, we were always asked to make up for schedule delays on the front end, and this left me feeling more than a little disgruntled. It all came to a head one afternoon.

I was sitting in my office, working on a project for a refinery in Ohio, when a sales engineer and his regional manager from a company called Beckman Instruments walked through my door. They were there to review a proposal from their company, but just as they sat down, my phone rang. It was my project manager. He had a list of complaints about what I was doing with the refinery project, and we engaged in a heated conversation.

Hanging up the phone, I turned to the two men in my office and said, "Hey, you don't need another salesperson at Beckman, do you?"

The regional manager paused for moment and said, "As a matter of fact, we do." He went on to tell me that Beckman was developing a line of measurement transmitters, controllers, and recorders with the goal of becoming a more complete automation company. As such, it was his job to build a team of control specialists as part of the launch. Before walking out the door, he handed me his card.

"Give me a call if you're interested," he said.

I thanked him and tucked his card into my wallet. A few days passed before I picked up the phone to give him a call. "I'm in," I said. And shortly after, my wife, Charlotte, and I packed up our belongings and moved to Houston. Was it a risk? Certainly. But I also knew Houston was a mecca for automation professionals and recognized that if this job at Beckman didn't work out, there would be many other opportunities available.

Still, this didn't stop the doubts from rolling in. *What are you doing, John?* I told myself. *You've got a young family to take care of. You can't just pull up roots and move to a new community. Also, you're an automation engineer, not an automation sales engineer.*

But in the fall of 1974, I pressed through my fears and made the leap.

KEEP ON PRESSING

I've always been a bit of a history buff. Give me a good World War II book on Winston Churchill or Harry Truman, and I'm engrossed for hours. What fascinates me about individuals like this is not only their strength to withstand immense pressure but also their ability to inspire a nation to press through periods of uncertainty.

In the closing words of his famous "Their Finest Hour" speech, Churchill delivered this powerful remark:

> Let us therefore brace ourselves to our duties, and so bear ourselves that, if the British Empire and its Commonwealth last for a thousand years, men will still say, "This was their finest hour."[13]

Note what he does with these words. He speaks to men and women in the present, asking them to step into the future and look back on their lives after they have made their decisions. *Will they look back with pride or with regret?*

---

13   Winston Churchill, "Their Finest Hour," The Churchill Centre, accessed August 4, 2023, https://winstonchurchill.org/resources/speeches/1940-the-finest-hour/their-finest-hour/.

On June 18, 1940, there was much reason to doubt Churchill's words. Hitler appeared unstoppable. And yet, Churchill acknowledged the severity of his nation's situation and inspired people to action. Harry Truman had a similar impact.

I admire Truman because it was never his plan to be the focal point of World War II. Truman was a businessman-turned-politician. As a vice president selected for his ability to broaden the Democratic Party ticket, Truman was far removed from most of the major decisions prior to Franklin D. Roosevelt's death. And then on April 12, 1945, everything changed, and suddenly Truman found himself in the global spotlight.

Prior to his inauguration, Truman didn't even know about the Manhattan Project. But less than four months later, he was the one who made the call to drop two atomic bombs on Hiroshima and Nagasaki—thus ending the US war with Japan.

I've gleaned a lot of wisdom from leaders like Churchill and Truman. And any time I was faced with a problem I thought insurmountable, I pictured leaders like them and asked myself what they would do. Sometimes the answer was to pivot and make a course correction. But always, the answer was to keep moving forward. It wasn't turning a blind eye to my doubts and fears. Instead, it was having the boldness to confront them and press through.

HERE WE GO!

For two years I served at Beckman as a Control System Sales Specialist, and my time there was a mixture of good and bad. In retrospect, Beckman's plans were more than a bit ambitious. They were in the middle of developing too many new products, many of which were not even close to prime time. As a result, the initial products we released were plagued with problems, and I learned the hard way how important quality is in automation production.

But once again, these painful learning experiences proved beneficial. Being a sales engineer exposed me to all kinds of other engineers, and I saw the whole gamut of automation professionals at work. These professionals were dedicated to their craft, and I learned a lot by calling on them. But after two years, my frustration resurfaced, and I knew it was time to make a change. In retrospect, some of my angst was impatience, but much of it was personal ambition. I wanted to grow at a faster pace than most company structures allowed.

My wife Charlotte sensed my dissatisfaction, and one Sunday, as she read the newspaper, she came across an ad for a company that said they were looking for a Director of Marketing. The company was unnamed, and Charlotte suggested I sign up for an interview session they were hosting at a local hotel.

After a brief call, I discovered the company's name was Rosemount—a small aerospace organization in Minnesota attempting to enter the process market. I'd heard this name in conversation before when my previous company reviewed one of their transmitters. And all I knew was that the little model 1151 transmitter we had passed around our office was now making major inroads in the market. Despite being a young company, Rosemount had a tremendous reputation. Every customer I'd spoken with that had tried Rosemount had nothing but praise for their organization.

Following my drive to the hotel for an interview, I met with the head of the Industrial Division. He was an impressive young man, and we connected right away. This first interview turned into a trip to Minnesota for more extensive interviews. Everything seemed positive. The leadership was strong, the people were wonderful, and it was obvious the company was headed in the right direction.

Still, after receiving an initial offer, I wasn't sure what to say. We had two small children, and I'd already asked my wife to support me in two moves. I told her I was going to turn it down. But Charlotte knew I liked Rosemount and wanted me to take the position. And this realization led to a memorable conversation. Being young parents, there were always children's books strewn about our home. Glancing at the floor, Charlotte picked one up, handed it to me, and grinned. The title was Richard Scarry's *Hop Aboard! Here We Go!*

Throughout our fifty-plus years of marriage, Charlotte has been my soulmate, best friend, adviser, and supporter. I couldn't have asked for a better spouse. And so, after a few more conversations and sleepless nights, we packed our bags and moved to Minnesota to begin what was the most wonderful ride I could have ever imagined. A journey that included its own fresh share of doubts.

THE POWER OF A GREAT CULTURE

On January 5, 1976, I reported for my first day of work at Rosemount. I remember it well because the temperature was just under 10 degrees. As I shuffled from the snow-packed parking lot toward the entrance, I wondered if I'd made a huge mistake in trading the sunny days in Houston for the frigid air of Minnesota.

But once I stepped through the doors, the welcome was warm, and I knew I'd made the right decision. There was so much positive energy, and everyone seemed excited. The mantra at Rosemount was "engineering excellence and innovation." And given the nature of the projects they tackled, excellence was exactly what was required.

The company began as an aerospace organization, designing products for the emerging jet aircraft industry and space program. In fact, it was Rosemount that supplied the oxygen flow sensor in Neil

Armstrong's backpack on the moon and the core temperature sensors that were to be embedded in the moon's surface.

I remember chatting with some of the employees who worked at Rosemount in 1969 and listening to them describe the night of the moonwalk. It was more than a little stressful. Most of the team was at the company watching the television, knowing that if their flow sensor failed, the first man on the moon would die. That experience made everyone understand the importance of that old saying, "quality first." And this same attention to detail they devoted to aerospace carried over into how they manufactured industrial products.

My title was Industry Marketing Manager for the Chemical and Petroleum Industry. By this point, Rosemount's pressure and temperature transmitters were starting to come into their own, and the company had a digital control system that was way ahead of its time. Keep in mind, this was the 1970s. There was no voice mail, no internet, no mobile phones, no fax machines, and no screens in the control room.

Control consisted of transmitters in the field sending back either analog electronic signals or pneumatic signals. And controllers in the control room were single loop analog with separate recorders next to them and an alarm box up above with back-lit panels to sound alarms. The control room panels were huge, and operators worked on their feet.

A typical early control room of the 1970s

For those of you who are more technical readers, the reason Rosemount was able to enjoy transmitter sales growth without being a major supplier of control room equipment was because by then the industry had basically standardized on the 4–20 milliamp current signal. The field transmitters simply varied the electrical current they sent to the control room between 4 and 20 milliamps in proportion to the range of the measurement they made. For example, if the measurement range was 0–100 gallons per minute, the transmitter sent 4 milliamps for 0 flow and 20 milliamps for 100 gallons per minute flow.

This meant Rosemount transmitters could be connected to *anyone's* control system—thus strengthening the company's reach in the marketplace. As a result, I spent my first few years at Rosemount supporting field sales and making suggestions for further development of the transmitters.

CONTINUING TO CLIMB

A few months after joining, Emerson acquired Rosemount. From the start, Emerson's plan was to use Rosemount as a linchpin to grow an industrial automation business. Chuck Knight was the CEO of Emerson, and our team at Rosemount was brought into Emerson's management process. In the back of our minds, we thought we could teach Emerson a thing or two. But as it turned out, we each had a lot to learn from the other.

During these years, a man I spent a great deal of time with was the president of Rosemount. His name was Vern Heath, and he was one of the company founders and among the first to create a profit-sharing retirement program—a concept far ahead of its time. Along with this program, Vern had other innovative approaches to leadership. He posted salary ranges for all to see and made all organizational job openings public. Vern held himself to a very high standard and expected the same from those he led.

I remember one time an executive in a hard-walled office brought in a radio. When Vern found out, he asked this man to take the radio home. In Vern's words, "[I]f our workers on the line can't bring in a radio, you can't either." This sense of equality made Vern a hero to all who worked with him. He had a way of making people feel special. Charlotte still remembers our first Rosemount Christmas party. We arrived a few minutes after the event started, and when we walked into the hotel ballroom, Vern was across the room chatting with a few of his friends.

As a childhood victim of polio, Vern walked with crutches. But as soon as he saw us, he hobbled over to where we were at and said to Charlotte, "I want to thank you for supporting John in his move up here. We're happy to have him." This act of thoughtfulness and total disregard for his own personal comfort epitomized the man Vern was. And my connection to him catapulted my career.

In 1978, I accompanied Vern on a trip to the People's Republic of China to put on a seminar at a hotel in Beijing.

People gathered to see Westerners wherever we went in China

Ever the champion of global business, Vern was excited about this opportunity. And because I understood Rosemount's products and was a good speaker, I was selected as a company presenter.

The whole trip was a bit surreal. After the State Department issued us special passports, our entire trip to China was tense. Our regular passports all had Taiwan visas, which we feared might bar us from entering. (China was not what it is today.) Our team was assigned a member of the Communist Party, and we were certain our rooms were bugged. When we landed in Beijing, the plane stopped on the tarmac close to the terminal. Dozens of armed Red Guard troops made a phalanx on each side, resulting in a fully guarded path off the plane's stairs and into customs. As we passed, the troops stood at attention.

In that moment, every member of our team had some self-doubt. Everything felt like a scene out of a movie. Vern was visibly frightened, and I remember coming up beside him and saying, "Don't worry, Mr. Heath—they didn't bring us over here to shoot us." Internally, I thought to myself, *at least I hope not!*

There is something about walking with others through difficult days. It forges closer friendships than peaceful times allow. And spending two weeks with Vern in an uncertain China allowed him to see another side of me he hadn't witnessed. He grew to trust me and my abilities. And soon after touching down in our return to Minnesota, my career skyrocketed. Over the next few years, I went from being a single industry marketing manager to the Director of Sales and Marketing.

I worked at Rosemount from 1978 to 1993, and after my role as a single industry marketing manager, I became director of Industrial Marketing in 1976, vice president of Sales and Marketing in 1980, and president of the Industrial Division and group president in 1989. By the early 1990s, I oversaw nearly three thousand employees. These years flew by and required a high level of personal sacrifice.

But I hoped it would all be worth it. In the back of my mind, I knew if I worked hard, it was only a matter of time until I had a good shot at the top position at Rosemount. And for fifteen years, my career trended in this direction.

THE GIANT OF SELF-DOUBT RETURNS

Given that I was close to the president of Rosemount and had a series of wins under my belt, I assumed it was only a matter of time until I assumed his position. My self-confidence was high. Vern was close to retirement, and I couldn't wait to take the company to new heights.

But then, everything came crashing down. In the spring of 1990, Vern called me into his office. After I sat down, he told me he had decided to retire. Now I knew why I was here. He had called me in to talk through his plan of succession, and he wanted me to be the first to know because he had selected me as the next leader. But quick as that thought entered my mind, Vern brought me back to reality.

"John, I'm announcing my replacement tomorrow," he said abruptly. "I'm sorry, but it isn't you." And two minutes later, the meeting was over. I didn't know what to say. All I could do was stagger out of his office and find some space to decompress. I couldn't believe what I'd just heard. There was no opportunity to interview for the position, no chance to share my vision, and no explanation for why I hadn't been selected.

In less time than it took to discuss an expense report, I was told my dream position wasn't even an option and that everything I thought I'd been working toward was no longer on the table.

That day was one of the lowest moments of my life. After a series of false starts in my career, the past fifteen years had been a meteoric rise as I had racked up one win after the next. I thought I had it all: the experience, the vision, and the skills necessary to lead at the highest level. But now, that old giant of self-doubt returned. *John, who have you been fooling the past few years? You thought you were big stuff. You thought you had it all together. But you really don't have all the qualities for a top leadership position. You never did.*

When I was new, I could chalk up missed opportunities to youth and inexperience. But now, I was at the top of my game, and my best wasn't enough. I was hurt, frustrated, and angry—mostly at myself. What had I done that caused me to be overlooked? Did I have some glaring weakness I'd never considered? To this day, these questions remain unanswered. Vern never did explain. And in some ways, this mysterious outcome serves as a powerful life lesson. Sometimes we can do everything we know to do and still come up short. We try our hardest and miss the promotion or get overlooked for a raise. When this happens, it's easy to let doubt set in and control our lives.

But when the giant of doubt rears its ugly head, there are several paths we can take. In my case, I could have become bitter and resentful and could have sought to undermine the new leader. Or, I could have wallowed in pity. But I determined not to live this way. I refused to burn any bridges and allow doubt to run my life. Instead, I focused on turning this giant of self-doubt and using it as fuel to make me a stronger leader. And in the months and years to come, I was grateful I made this decision.

FISHER AND ROSEMOUNT MERGE

In retrospect, not receiving the president's job at Rosemount turned out to be a blessing in disguise. In three years, Emerson (who you'll recall owned Rosemount) acquired a company called Fisher with the intended goal of merging this company with my current place of employment, thus creating one major organization—Fisher-Rosemount. To understand the significance of this merger, it's important to back up and offer a brief, albeit technical history.

Rosemount began as an aerospace company and diversified into process transmitters in the late 1960s. Their approach was truly disruptive because their transmitters had no moving parts and were extremely rugged and stable. As a result, they grew quickly. When Emerson wanted to get into the world of process devices, they acquired Rosemount in 1976, and Rosemount soon became the anchor company in Emerson's strategy. As time passed, additional measurement devices were acquired along the way and placed under Rosemount's wing. By the late 1980s, Rosemount was the largest process measurement company in the industry.

But measurements alone were not enough to control operations in a plant. They were the eyes and ears. But there needed to be a brain and hands to manipulate the flows through the pipes in the plant to bring and hold various parameters to their desired value. This was the strength of Fisher.

Fisher control valves on the right

Like Rosemount, the history was fascinating. Fisher was an older company, founded in Marshalltown, Iowa in 1880 when a bicycle shop owner named William Fisher came up with a revolutionary idea. As a volunteer fireman, Fisher grew tired of the lack of constant pressure in fire hoses. He knew there must be a better way, and so he developed a device called the "Fisher governor." And within a few years, Fisher became the world's largest supplier of control valves and regulators in the world.

To understand what a control valve does in simple terms, think about a valve in your shower and how you adjust it to get the temperature you want. When you do, you get in and hope the water pressure and temperature remain the same. The valve is in one position, and if the supply of hot or cold water changes temperature or pressure, you're in for an unpleasant surprise. Take this concept and provide a valve with a remotely controlled actuator (the hands), and you have an idea of the value Fisher brought to the market.

Now comes the brains of the operation, and this is where both Rosemount and Fisher enjoyed the advantages of standardized communication between the field and the control room. In the early 1970s, the whole control industry adopted a standard way to communicate from the control room to the field. Measurement devices were each connected by a pair of dedicated wires to the control system. And the control system compared the actual value of the measurement to what was needed to make the product in the plant.

From there, the control system then sent a signal to each control valve over a different dedicated pair of wires. This meant a customer could buy a Rosemount measurement device and hook it up to any control system. They could also buy Fisher control valves and hook them up to any control system. A standard communication protocol enabled customers to buy a measurement product from one supplier, a control system from any control system supplier, and a control valve from any control valve supplier, as long as each of the suppliers conformed to the standard.

When Emerson acquired Fisher in the fall of 1992, this acquisition gave Emerson another leg of the automation stool. The leading control valve company (Fisher), combined with the leading measurement company (Rosemount), gave Emerson tremendous presence in front of the customer. It also offered Emerson several areas of synergy in technology, sales, and marketing. Both Fisher and Rosemount had small control system businesses, which were not major players in the control system market, and these two businesses were merged into one company called Fisher-Rosemount Systems. The Rosemount system business was in Minnesota, and the Fisher system business was in Austin.

WHAT HAVE I GOT TO LOSE?

When Fisher and Rosemount merged, I was a group president at Rosemount and suddenly found myself out of a job. Acquisitions often creates an odd person out, and I was that person. With a cloud of uncertainty hovering overhead, I started to interview at other organizations. In my mind, I wasn't needed anymore, and I had to look after my family.

But after a few months, the newly appointed overall business leader of the combined Fisher-Rosemount, Joe Adorjan, came to me and asked if I would move to Austin and lead the confluence of the two small control system businesses, now named Fisher-Rosemount Systems.

Combined, the system businesses of Fisher and Rosemount were losing $40 million a year. The big and profitable measurement side of Rosemount looked at the system business as a financial black hole. And the control valve side of the big and profitable Fisher felt the same way. Emerson wanted measurement and control valves. But they didn't care much about the system business.

In Joe's words, I was to go down to Austin and prevent the system business from "F***ing us up too badly." Not exactly a brilliant motivational strategy, but I understood his expectations. I was not to be an innovator. Instead, I was to stay in my lane and hold everything together until the systems business could be sold or shut down altogether.

In 1993, I was introduced to a packed room of employees in Austin as the president of Fisher-Rosemount Systems. The date of this announcement? April 1. To many in the audience, this news must have felt like a feeble attempt at an April Fools' prank. But as I rose to speak, I assured everyone present that this was real.

The tension was palpable. Everyone was in shock, and I knew what many were thinking. *Here is this guy from Minnesota coming to shut us down or sell us on a vision that is doomed to fail.* Because we were merging two actual competitors, this meant many roles were redundant. And since I had spent the past fifteen years at Rosemount, many assumed I would hire all my buddies in Minnesota and kick people in Austin to the curb. It was clear I would have to earn their trust, and doing so would take time.

I'll detail some of the toughest obstacles I faced in the next chapters, but let's just say that tackling this delicate situation proved more difficult than I could have ever imagined. I'd be lying if I said there weren't moments I doubted my ability to bring these two organizations together. Some days, I felt like I was hanging by an emotional thread. We'd gain traction, seem to make progress, and then *wham!* we'd get hit with an unexpected challenge.

One of these moments came about a year into this merger. I'd asked the head of operations at Rosemount to move to Austin and oversee all joint operations. I knew this person well and was confident he would do a great job. And he did. Soon after settling into his new role, his department made great strides. But about a year into his tenure, he walked into my office and shut the door.

"John," he said, "I'm sorry to do this to you. But I've decided to move back to Minnesota with another company. It's nothing against you, but that city is a better fit for my family."

I nodded, too stunned to say anything. As soon as he left, I put my head down on my desk and wept. It was the first time I recall ever doing this at work. All I could think to myself was, *I cannot handle this. I'm not cut out to be a leader. It's too painful, and the work is too demanding.* I still think back to this day as being one of the lowest moments of my life. That evening I went home and knew I had a choice to make. I could let this disappointment eat me alive, or I could rise above it.

While I have a hard time explaining this to some, whenever I'm faced with a serious challenge, I sometimes have what could be described as an out-of-body experience. It's as if I'll hover above my problems and look down on them. And doing so brings me clarity.

Sometimes when we're so immersed in the pain and doubt of a certain situation, it's tough to see a path forward. But I've found that when I'm able to step back and look down on my problems, I'm able to make the next logical move. I free myself from emotional attachment and focus on what matters most.

It's a bit like getting lost in a forest and scaling a tree to regain our bearings. When we're on the ground, what we see isn't pretty. But 40 feet in the air provides perspective, and we see our location in the context of where we need to go. Our goal then becomes to take one step forward in the right direction, and then another, and then another. And pretty soon, we're back where we need to be.

BE THE TOY CLOWN

Before I close this chapter and move on to the giant of innovation, I need to make a final point about the role of personal sacrifice and the internal doubts one must confront away from the office.

It's easy for some to look at my résumé and think, *John, you've had a great life*. And make no mistake, I have. But I'd be remiss if I didn't point out the many internal battles I've fought along the way.

For all those years leading up to the Fisher-Rosemount merger, I'd paid a steep personal price for the amount of sacrifice and personal dedication these jobs required. I spent nearly 70 percent of my workdays out of town and missed countless nights with Charlotte and the children. And after the merger happened and I had my new title, this level of sacrifice increased.

It was about a year after I'd made that fateful April Fools' speech to a packed room of dubious employees that I confronted the greatest giant of doubt I'd ever faced. And this was the giant in my mind. Today, it's much more acceptable to throw around terms like "personal anxiety" or "depression." But in the mid-'90s, this was not even a point of consideration. And rather than reach out for help, I did the only thing I knew to do and doubled down on my work. As a result, my body paid the price. I lost my appetite and ate little more than a piece of toast for supper. My stomach always ached, and I lost a lot of weight. The stress of bringing two companies together was overwhelming. I felt like I was on all the time and could never flip the switch, even when I came home.

Work was stressful, but life at home was not any easier. My family was sad that we'd relocated from Minnesota to Austin and didn't know anyone. And while I was excited to be in my new role at work, Charlotte felt isolated. While I did my best to hold the company together, she was the one tasked with finding new doctors, new dentists, a new job, and new friends. More than once, I had doubts about whether I'd made the right call for our family.

But as was the case so many times before, time proved the best healer of doubts and pain. On my darkest days, I often thought about a toy clown I had when I was growing up. It was a silly little inflatable clown that had a base filled with sand.

And because of the way it was balanced, my ten-year-old self could punch the clown squarely in the nose, causing the clown's head to hit the floor, only to have it pop back up into place. It didn't matter how hard or how many times I hit.

It's funny how different memories from your childhood stick, and this was certainly one of them. When life hit me hard and doubts emerged, I was determined to be like the clown and get back up. I knew difficult days would not last forever. In the words of Ralph Waldo Emerson:

> Finish every day and be done with it. For manners and for wise living it is a vice to remember. You have done what you could; some blunders and absurdities no doubt crept in; forget them as soon as you can. Tomorrow is a new day; you shall begin it well and serenely, and with too high a spirit to be cumbered with your old nonsense. This day for all that is good and fair. It is too dear, with its hopes and invitations, to waste a moment on the rotten yesterdays.[14]

Through tough times, I've learned the value of breaking large tasks into smaller chunks and tackling each challenge one at a time. This is how I turned the giant of doubt. I've realized that everyone has doubts about themselves. It's part of being human. In fact, there is a word for someone with zero self-doubt—dictator! Some doubt is good

---

14   James Elliot Cabot, "Chapter 13: His Ways With His Children," in *A Memoir of Ralph Waldo Emerson*, Volume 2 of 2 (London and New York: Macmillan and Company, 1887), Quote Page 106 and 107.

because it keeps you on your toes. And if you leverage it correctly, you can use your doubts as fuel to propel you forward.

Turning the giant of doubt isn't easy. But when you're faced with a setback that increases your doubt, here is what I'd challenge you to do. First, try to float above your situation and look down on your problems. When you do, they're probably not as bad as you think.

Next, break your greatest doubts into smaller bites, and tackle them one at a time. Be like Churchill, and imagine yourself in a few months or years from now. *How will you have wanted to have spent this time?* Don't live your years with regret.

Finally, keep pressing forward. Don't quit. In the words of Yoda of Star Wars, "Do or do not. There is no try." Confront your doubts. Name them. And then turn them.

CHAPTER 3:

THE GIANT OF INNOVATION

After I took over as president of Fisher-Rosemount Systems, I faced a tension between the mission I was given and what I knew I needed to do. I was hired to plug holes in our leaky financial ship. But I knew the greatest giant our new company faced was the giant of innovation.

If I turned our organizational finances around, but our competitors out-innovated us, it wouldn't matter what kinds of profits we made because we'd soon find ourselves with a shrinking top line. The giant of innovation is tricky.

On the surface, innovation is simple. In the words of Tom Freston, "Innovation is taking two things that exist and putting them together in a new way." It's not necessarily coming up with a new system from scratch, but it's putting all the pieces of a puzzle together in a way that others haven't.

On a global scale, I think of countries like Japan and China. Following World War II, the Japanese took a very innovative approach to production. But as the decades passed, they've since fallen behind on innovation. Sony invented the Walkman but totally missed smartphones. Today, China is a dominant player because the Chinese have a clear vision of where they want to go. There isn't near the level of bureaucratic red tape that companies in America face today. China doesn't play by the same rules Western nations do. They are clearly trying to be a global disrupter, not through innovation but through sheer force of will.

I recall reading *The Innovator's Dilemma* by Clayton Christensen. And in this book, Christensen demonstrates why great companies fail. He writes, "[T]here is something about the way decisions get made in successful organizations that sows the seeds of eventual failure."[15] According to Christensen, the way some companies avoided this trap was through "disruptive innovation." In his words,

> Disruptive technologies bring to a market a very different value proposition than had been available previously. Generally, disruptive technologies underperform established products in mainstream markets. But they have other features that a few fringe (and generally new) customers value. Products based on disruptive technologies are typically cheaper, simpler, smaller, and, frequently, more convenient to use.[16]

As I surveyed the competitive landscape in 1993, I knew Fisher-Rosemount Systems had only one real option on the table. We had to innovate. We had to disrupt. And if we didn't, it was only a matter of time until our new company would cease to exist.

---

15   Clayton M. Christensen, *The Innovator's Dilemma: Management of Innovation and Change* (Boston: Harvard Business Review Press, 1997), 20.

16   Ibid., p. 25.

To understand why I knew this to be the case, we need to back up a few years and lay a proper framework. Some of it is a bit technical, but stick with me.

THE HONEYWELL GIANT

In the early 1970s, there were two leaders in the control business—Foxboro and Taylor. Control in those days was not a "system." Rather, control was a series of individual boxes mounted in a panel. These boxes had meters on the front that displayed the measured variable and had a dial with which the operator entered the desired value of the variable. (I told you this was technical.)

Each stand-alone controller dealt with only one measurement and had analog circuitry to compare the actual value of the measurement to the desired value. From there, the controller would generate an output signal to the control valve to bring and hold the measured variable to the desired value. Each panel had numerous controllers mounted, and the panel was quite large.

Old Control Panel

Enter Honeywell.

Today, you might be familiar with Honeywell products if you have a thermostat in your home. But in 1976, Honeywell disrupted the control business by introducing a microprocessor-based control system. At the time, Honeywell had stand-alone controllers, but they had always been a minor player in control.

But now, with the emergence of this new product, this microprocessor ran fast enough to do multiple loops at a time in one controller. And there were these proprietary operator stations with keyboards and cathode ray tube (CRT; tube TV screen) displays. To some who were more conservative, this breakthrough was right out of a science fiction novel. No one believed microprocessors were ready for prime time, and there was great skepticism that operators could work off a glorified TV screen.

However, over the course of the next few years, Honeywell made inroads into the control business. They named this new system TDC 2000 and coined a phrase to describe it—a distributed control system (DCS). Honeywell became the leading supplier of control systems, and this disruptive technology forced all single loop suppliers to rush into distributed control systems.

Simply put, Honeywell changed the game.

As a result, Honeywell enjoyed first mover advantage and built a massive installed base. Why was this disruption important to the measurement transmitters and control valves? At first it wasn't because TDC 2000 accepted the standard analog signal to and from the transmitters and valves. But here was the catch.

Despite also having a line of measurement transmitters, Honeywell could never make a dent in Rosemount's business. While the market was accepting microprocessors in control systems, it was still leery of putting them into field devices because the field is a hostile environment compared to the climate-controlled control room.

Typical hostile plant environment

Once again, Honeywell was bold, and in the late 1980s, they introduced a smart, microprocessor-based line of measurement transmitters. They used the processor to improve accuracy in the transmitter, but they also used it to introduce a proprietary, digital way of communicating with their control systems. Practically speaking, this meant there was still a pair of wires going from each measurement transmitter to the control system, but those same wires could now be used to send digital signals, which offered certain advantages to customers.

One of these advantages of digital communication was customers could now use a handheld device that looked like a calculator to "talk" to the transmitter. This dramatically simplified the setup process. "Dumb" transmitters had to be calibrated by turning screwdriver adjustments in the transmitter. But "smart" transmitters could be easily set up from almost anywhere along the wire. And over time, even Honeywell's control system could emulate what handheld devices accomplished. The catch was that the digital way the Honeywell transmitter communicated was proprietary to Honeywell and would only work with their system.

Honeywell's strategy was clear. Because they were at the top of the ladder with control systems and near the bottom of the ladder in transmitters, why not find a way to leverage their control system strength?

When I first saw Honeywell's smart transmitter, I cursed, realizing they had changed the game. If their strategy became a trend, all big control suppliers would develop their own proprietary field communication. And the result could be a digital Tower of Babel, with Rosemount and Fisher winding up tongue-tied.

A modern comparison today might be Apple. Steve Jobs always talked about Apple creating a "walled garden." Because of this concept, there are certain programs that only run on a Mac or IOS device. When one buys into Apple, one buys into their ecosystem. As a result, competitors like Microsoft and Samsung might develop a superior laptop or smartphone, but because Apple controls the hardware *and* the software, they have a very strong advantage with consumers who want continuity between all their devices.

Honeywell was trying to squash the competition, and this meant they were trying to squash people like me!

I immediately called an emergency meeting. We decided to scrap our existing development programs and go all in on smart devices, even though they could cannibalize our existing products. We would also fight proprietary protocols by developing a superior protocol that allowed for simultaneous analog and digital communication over the same pair of wires.

Then, we would donate the protocol technology to the industry by forming a foundation to manage it. And from here we would work to get other companies to join our efforts. Customers didn't really want the Tower of Babel. They wanted the same kind of standardization they had in the pure analog world.

The lesson learned here is that if you are on top and a disrupter comes along, act quickly and decisively, even if you risk cannibalizing your current products. If you don't, the disrupter will consume your current products.

MY GREAT DILEMMA

I understood Honeywell's endgame, so when I assumed the role as president of Fisher-Rosemount Systems, I had a clear objective in mind. While the management at Emerson wanted me to play defense and minimize our losses, I realized we needed to do the opposite.

With this understanding in place, I reached out to a friend of mine named Bud Keyes. Bud was a respected person in the industry who would later be inducted into the Process Control Hall of Fame. Bud had served as the president of one of our competitors but had since shifted into a consulting role. When Bud spoke, people listened. He's the closest thing to a genius I've ever met.

Bud was the type of guy who only needed three hours of sleep a night. He was a master in every conversation. In the morning, he could sit in a room with the best and brightest tech experts and speak their language. Over lunch, he could walk in a room with business-people and talk P&L and market projections. And in the evening, he could sit around chatting with guests at a cocktail party about the literary greats of the past. He was the ultimate renaissance man.

Initially, all I wanted Bud to do was provide an honest evaluation. Fisher's and Rosemount's System businesses were now one, but they entered this merger with separate visions for the future. And I wanted Bud to evaluate what he thought of each company's approach.

After a few days of interviews, Bud handed me a detailed report and confirmed my greatest suspicions. Prior to the merger, both Fisher and Rosemount were focused on *keeping pace* with the control systems competition and were not anticipating *new* innovations. They weren't factoring in what was going on outside our industry in the PC, networking, and commercial technology world.

Bud agreed that the leading market position from Honeywell was real and growing. And based on this reality, he recommended we make a bold, disruptive move. After spending some time with members of our Fisher-Rosemount Systems team, he thought we had just the people to make it happen. Still, it would take a lot of work, and the odds of success weren't exactly in our favor. There was so much that could go wrong.

After reviewing Bud's report, in the fall of 1993, I took Bud to St. Louis to see my direct boss, Joe Adorjan. As the business leader behind the Fisher-Rosemount Systems merger, I knew the only way we'd be able to move forward was if Joe was on board. Over the course of an hour-long conversation, Bud outlined for Joe what he'd already shared with me. Fisher-Rosemount Systems couldn't afford to sit back and play conservative. We had to be innovative and swing for the fences.

After Bud's impressive presentation, I turned to Joe and said, "[W]e really need to do this."

There was no doubt in my mind that if Fisher-Rosemount Systems made some bold yet calculated decisions, we could give competitors such as Honeywell a dose of the innovator's dilemma. Clayton Christensen's book made me realize that Fisher-Rosemount Systems needed to solve the customer's pain points through leveraging some of the emerging commercial technology. If we did, we would be a disrupter.

Of course, if I were to be perfectly honest, I'll have to admit there was some personal ego involved. After getting overlooked for the presidency at Rosemount, I wanted to prove to myself and others that I could be more than a caretaker. But more than ego, there were some core values that motivated my behavior. One of them was a holistic view of automation.

A HOLISTIC VIEW OF AUTOMATION

While Fisher-Rosemount Systems was a subsidiary of Emerson, this didn't limit my view on how Emerson should function. Essentially, there were three legs of automation: *measurement products (Rosemount), control valve products (Fisher), and systems products (Fisher-Rosemount Systems)*. Emerson had their hands on all three.

However, Emerson was only making money in measurement and control valve products. And as president of the systems business, I sensed this tension. Many at Emerson wanted to kill the systems business altogether. After all, why be involved in a business that was barely over $200 million in sales and lost $40 million a year?

But even before assuming my new role as head of systems, I was convinced it was in Emerson's best interest to remain engaged in all three parts of automation. The reason was simple. If the three pieces worked together, Emerson could deliver entire projects to customers from start to finish. And this would give us a strong competitive advantage.

I knew what we needed to do to compete. We needed to be strong in the system business. If our measurement transmitters, control valves, and control systems all worked with a common "look and feel," we could offer a whole new layer of customer advantages. We could use the digital communication standard to deliver strong diagnostics, which would enable customers to better manage their physical assets in the plant.

Along with this, we also needed to push communication standardization so that the Rosemounts and Fishers of the world could sell products not only connected to *our* system but also connected to *any* company's system. After all, no customer wanted to deal with overbearing proprietary obstacles.

TAKING MY IDEA TO HEADQUARTERS

After my meeting with Joe, I was more convinced than ever that we were on the right track. Joe was as well. In fact, he was so impressed with Bud's presentation that he hired him several months later for a position in St. Louis. And in the coming years, Bud served as a sort of consultant to Fisher-Rosemount Systems.

Now that Joe was on board, it was time to take my idea to the CEO of Emerson, Chuck Knight. Thankfully, I wouldn't have to wait long to get my chance. Once a year, presidents and key employees of various companies tied to Emerson traveled to St. Louis for an all-day meeting with Chuck and other senior-level Emerson executives. Presenters would speak with two hundred to three hundred PowerPoint slides in an intense, Super Bowl-like atmosphere. This was our time to flop or shine.

Because I'd been with Rosemount for over seventeen years, I understood the full significance of this meeting. The first time I ever presented to Chuck Knight was in 1979 when I was the head of marketing at Rosemount. I started my presentation with a series of projections for the future of our company. And while I can't remember all the details, the first half hour went great. Chuck seemed engaged, and I was just finding my stride.

However, forty-five minutes in and my presentation went off the rails. I got to the part of my talk where I projected that the margins of Rosemount would trend down in the next few years. From my perspective, this made logical sense when looking at the larger picture. Sure, this would make it tighter financially in the short term, but I knew we would come out well ahead in the long term. We needed to expand globally, and this required up-front investment.

But Chuck didn't see it this way. As soon as I got to this point in my presentation, Chuck's head popped up, his expression changed, and he leapt to his feet. "God damn it, John," he said, "I believed everything you said until now. Do this and you'll drive our margins into the dirt. What are you thinking?"

I backed up, stammered slightly, and did my best to explain why I thought this approach was in the best interest of our company. To grow our business in the big picture, we had to invest a little more upfront. Chuck wasn't buying it, and he pressed me again, his eyes narrowing. Again, I did my best to explain and sputtered out a few additional facts to back up my point.

After listening to me for a few more minutes, Chuck folded his arms, sat down, and firmly stated, "Next."

I sat down, my brain in a panic. I thought I was done. In my mind, I had just gone through a career-limiting experience or worse.

But that evening shortly before our company dinner, I witnessed a different side of Chuck. We were drinking some cocktails, and I saw an opportunity to save myself. Still a little rattled from our previous exchange, I said, "Listen, I don't think I adequately explained what I think we should do. I'm happy to come back any time and outline my plan in greater detail."

Chuck smiled, put his arm around me, and said, "Don't worry, John, you did fine. I was just testing you."

Instantly, I felt relieved. It wasn't that Chuck didn't believe in what I was saying. He was just checking to see if *I* believed it. The lesson learned here is that it is a leader's job to test commitment. But they should do so in a way that challenges ideas and is not personal. Chuck was a master at this.

Over the next ten years, I watched as Chuck tested me and countless other presenters. All Chuck ever wanted was for his leaders to understand what they were talking about, have a clear plan of action, and believe in what they were doing. And whenever presenters failed on one of these three fronts, this became their real career-limiting move.

I share this because by the time I presented my systems ideas to Chuck in 1994, I'd already presented in front of him over twenty times. And in retrospect, it's a good thing I had, because the ideas I was about to share with him were radical and, dare I say, innovative.

EIGHT INNOVATIVE IDEAS

With this knowledge in the back of my mind, I entered Emerson's conference center in early 1994 with this mixture of excitement and uncertainty. In my mind, I knew the ideas Bud and I had shared with Joe would work. But would Chuck see it the same way? What would the other executives in the room think? I knew what they probably expected.

They assumed I'd walk in, and as the new president of a new company, I'd pitch some conservative projections and mumble something about reducing costs and increasing profitability.

And while I did present some ideas about improving financial performance, the centerpiece of my presentation was eight innovative ideas that would form the basis for a plan later termed "DeltaV." It was my belief these ideas would disrupt the control system business and offer us the inside track with customers.

Here is the gist of my outline from that day.

INNOVATIVE IDEA #1 | WE WOULD RUN OUR NEW OPERATOR STATIONS ON WINDOWS PCS

At the time, all control systems from competitors used custom-built, proprietary operator stations, which sold for $40,000 per screen. Contrast this with a Windows PC, which was around $2,000. To make this happen, we would partner with Dell. And Dell would then preload the PCs with Windows and our operator interface software. The hardware and software prices were decoupled, which would allow us to charge for the software and offer a subscription service.

Today, this idea sounds like a no-brainer, but in 1994, it was a big deal. Because Microsoft was still in their early stages of development, there were glitches that sometimes made their software unreliable. This posed a massive challenge for companies that needed their operator stations to function without any hiccups.

But here is what was clear. While Windows software and PCs were still a bit less reliable than proprietary operator stations, Windows would only get better with time, and PCs would get more powerful and less expensive. And as they did, the costs for consumers would drop and performance would increase. There was also the added perk that customers would have a Windows PC that would allow them to run other Windows applications, such as Excel.

INNOVATIVE IDEA #2 | WE WOULD DESIGN OUR CONTROLLERS AND OTHER RACK-MOUNTED HARDWARE TO TAKE UP A SIGNIFICANTLY SMALLER FOOTPRINT

As I shared with Chuck and the Emerson team, we would take advantage of the growing power of microprocessors and application-specific integrated circuits (ASICs).

While ASIC sounds like something from Star Wars, it is nothing more than a microchip designed for a special application. And the end result is it enables a whole circuit board to be shrunk into one chip. Today, these are everywhere.

Every automobile has a chip that controls the engine. And without these chips, the vehicle is useless. And when there is a chip shortage, as we've seen in recent years, entire car lots can sit full of vehicles that look intact but are missing one or two critical components.

Our goal was to shrink the whole system. And this meant a much smaller control house would be needed. This would prove especially helpful for customers that produced sensitive products. Since control houses in refineries and chemical plants needed to be blast proof, reducing the entire size could save a company hundreds of thousands of dollars.

INNOVATIVE IDEA #3 | WE WOULD USE ETHERNET FOR THE "DATA HIGHWAY" WITHIN THE CONTROL ROOM

Again, this sounds like a "no duh" idea. But in 1994, all control systems used a proprietary network for communication between and among controllers and for communication from controllers to operator stations. This data highway did not go out to the field and remained in the control room.

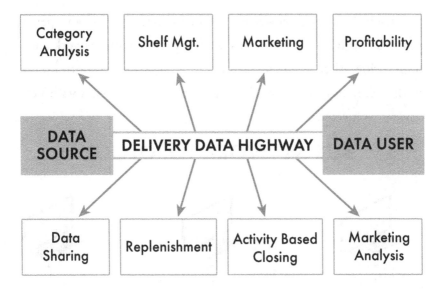

Because ethernet was an evolving commercial technology, I was confident it would get faster and more powerful at a lower cost. At that time, the cost for a proprietary communication board was $1,500. Contrast this with an ethernet setup that would only cost a tenth of this price. (Fun note: Ethernet is now just a chip that costs a few dollars.)

I should also point out an ethernet system was much easier to install.

INNOVATIVE IDEA #4 | WE WOULD BUILD OUR SYSTEM FOR BUSES TO COMMUNICATE WITH FIELD DEVICES

In case you're unfamiliar with the term, a bus is a name for a network with a main channel and several devices connected simultaneously. A simple analogy would be a surge protector for your home computer. There is one end that plugs into the electrical outlet, and inside the surge protector is a main channel that distributes power to the multiple outlets on the protector. And from this simple power bar, you can connect a PC, printer, monitor, lamp, and whatever else you need.

Each device gets power via the main channel and can operate simultaneously. A bus does the same thing, only with network communication. There is a main channel running from the control room to the field, and this channel can have multiple branches with field devices connected.

BUS NETWORK TOPOLOGY

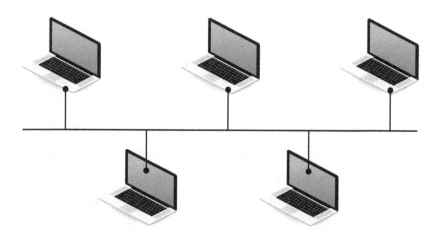

Each device has access to the main channel. This is important because it frees the customer from running an individual pair of wires from the control room to each field device, again resulting in considerable savings. We also anticipated the emergence of wireless communication between the field and the control room so that we could easily accommodate that in the future.

INNOVATIVE IDEA #5 | *WE WOULD ELIMINATE THE NEED FOR A FACTORY ACCEPTANCE TEST*

In the mid-1990s, every control system had to have a factory acceptance test (also known as FAT). This meant everything that was going to the customer's control room had to first be assembled at the supplier—including all cabinets, racks, wiring, and operator stations. Customers then sent a team of engineers to do the acceptance test, eating up a lot of time in the process.

After everyone agreed the system was functioning as it should, the customers went home, and the whole system had to be disassembled, packaged, and shipped to the customer's site where it was then reassembled. This entire process was a huge additional cost for both suppliers and customers. But because our system would be designed in a more modular fashion, we could ship the various parts of the system directly to the job site and assemble it there—thus eliminating the FAT.

Also, because DeltaV would be software based, we could offer "virtualization." This meant we could send the customer software and they could simulate the whole system on a PC. As a result, this new way of doing business would make it much easier for us and for customers.

INNOVATIVE IDEA #6 | WE WOULD BUILD A COMPLETE PLATFORM THAT WOULD GET BETTER WITH TIME

At the time, control systems were hard to upgrade. They functioned and grew outdated, and customers were faced with the decision to fork out money for a new system or get by with what they had. Our system would be very different and allowed for continuous innovation.

This meant the system customers bought on day one could evolve with us as we upgraded and rolled out new functionality. They would not need to "rip and replace." Our system was modular and kept track of the version of every piece of hardware and software. And it would allow us to work with our customers to develop a sensible and cost-effective migration plan over the years.

Time would prove this concept valuable because a DeltaV purchased in the early days can still be humming today. Controllers have changed, software has changed, networks have switched, but DeltaV rolls along unphased. Fisher-Rosemount Systems continued to introduce more innovation on top of the original platform, which kept us and customers current with technological advancement.

INNOVATIVE IDEA #7 | *WE WOULD OFFER REMOTE TROUBLESHOOTING*

So long as the customer agreed, Fisher-Rosemount Systems' home office experts could connect to their system and remotely troubleshoot any problems they faced. Sometimes a customer's issue might be as simple as entering an incorrect configuration. If this happened, we could identify this challenge remotely and fix it.

Once again, this concept of remote assistance is common today. But in the 1990s, this idea was far ahead of its era. And the result would save customers a lot of extra time and money spent on engineers.

INNOVATIVE IDEA #8 | *WE WOULD SELF-FUND THIS PROJECT AND IMPROVE PROFITABILITY*

Because I knew Chuck wouldn't agree to this idea if it meant Emerson lost even more money, I outlined how this entire project could be funded by Fisher-Rosemount Systems and improve our bottom line every year.

Despite any objections Chuck and other executives might have, I also knew that continuous improvement was in Emerson's DNA. And so when I presented to him, I followed the playbook. With a little bit of luck and well-timed business deals, we were able to land some significant new business and keep our eyes set on innovation. We didn't dig ourselves into a financial hole, and we became innovative disrupters.

In summary, compared to the systems that existed at the time, this new approach would be like laying down a royal flush compared to a small pair.

After I presented, Chuck and the other Emerson executives were out of breath, and so was I. We all recognized this was a bit of a moonshot. There was no instant approval, and in subsequent months, I had to present this same concept over and over.

CREATING A TEAM

When I left the conference, even though I didn't have full approval, I decided to move forward. As the saying goes, sometimes it's better to ask for forgiveness and not permission.

First, I knew I needed a strong leader to oversee the whole development project. As it turned out, Jim Hoffmaster was this guy. Jim had what I consider one of the most important traits in a great project leader—focus. When Jim set a goal, he zeroed in on it like a laser and would say no to any other emerging ideas that veered from the overall mission.

This is an important point to make. Unfortunately, many projects go off the rails because of mission creep. But with Jim at the wheel, I knew this would not happen.

After we brought Jim on board, we created a team of the best people in technology, marketing, and operations. And when these wheels were set in motion and people were assigned to work under Jim, I moved the entire team off-site and called the project "Hawk."

It was a small team of fifty people. They had relaxed dress codes and flexible hours. Today, you'll get laughed out of the building if you show up to a Silicon Valley tech startup in business attire. But in that day, relaxed dress codes were not in style. However, my goal was productivity. I didn't care what my engineers wore. I just wanted them to produce results.

My edict to Jim and the team was to remain focused on their mission. Under no circumstances were they to pull back from working off-site and get re-entangled in the day-to-day operations of Fisher-Rosemount Systems. I wanted them to innovate and knew this wouldn't happen if most of their time was spent on "sustaining engineering."

AN ENDORSEMENT FROM THE KING OF INNOVATION

In the summer of 1995, I got connected with a man named Mike Maples. At the time, Mike worked at Microsoft inside the Office of the President as the executive vice president of the Worldwide Products Group.

Mike was in town helping former president, George H.W. Bush, set up his personal computer (one of the perks of being an ex-president). And because I knew Mike was in town, I asked if our team could meet with him. He agreed to have lunch, and I gave him a similar, albeit shorter, pitch that I gave to Chuck.

I told him Microsoft needed to pay a little more attention to automation and that we would really like their endorsement. A few weeks later, we received this video from the king of innovation, Bill Gates.

> Hi, I'm Bill Gates. A core part of Microsoft's business is to work closely with industry leaders who share our vision. Today, I want to describe how Microsoft and Emerson Electrics Fisher Rosemount are delivering on a shared vision for process manufacturing. Manufacturers in the process industries have unique and demanding requirements for information technology.

Their competitive environment requires faster responses, improved productivity, and better management of plant assets. To really thrive in this environment, they need a seamless flow of information from the plant floor to users across the enterprise. New technologies from Microsoft and Fisher Rosemount are meeting these challenges. At the forefront are Microsoft's 32 bit clients and Microsoft Windows NT on the server.

And Fisher Rosemount Systems is building breakthrough applications with key Microsoft technologies like ActiveX, Olay for Process control, and the Microsoft Foundation Class Libraries. The resulting scalable process system technologies provide users with intuitive operation, outstanding performance, and true interoperability with desktop applications. In terms of new technologies on the plant floor, Fisher Rosemount is providing increased intelligence in field devices like valves and transmitters.

In effect, these microprocessor-based field devices are data servers in their own right. They provide a new richness of information about plant performance by providing data about both the process itself and the equipment that runs the process. These technologies will redefine process management. The game will change as integrated modular software applications emerge and take advantage of this new information. New suites of software targeted at plant applications like asset management will provide value to manufacturers far beyond process control alone. That's why I'm excited about the way Fisher Rosemount is using Microsoft technologies in a series of breakthrough products.

This outside confirmation of our work provided a huge boost to our efforts. It validated what we were trying to do and sent a message to customers and competitors that we were here to stay.

THE KEYS TO DISRUPTIVE INNOVATION

It was the French novelist Marcel Proust who said, "The real voyage of discovery consists, not in seeking new landscapes, but in having new eyes."

This is the key to disruptive innovation. When we see the world, or our marketplace, with new eyes, we understand where we need to go. And the best eyesight comes from getting into the minds of our customers. It's sitting with them, asking questions, and understanding life from their perspective.

Note that in every one of the eight innovative ideas I offered, each was focused on the customer. Offering a Windows operator station saved time, money, and space. Building an ethernet data highway and designing for buses did the same. Eliminating the need for a factory acceptance test, offering a complete platform, providing remote troubleshooting, and reducing footprint benefited consumers in tangible ways.

Obviously, this also benefited Fisher-Rosemount Systems, but starting with the customer in mind provided the proper "why" for our innovation. It is a solution looking for a technology, not a technology looking for a solution. In my mind, if we could make our customers' lives easier through offering a superior product and service, our largest problems were already solved.

As you've read this chapter, perhaps you've identified a few areas in your company where you need to innovate. You've thought about becoming a disrupter, but something is holding you back. If this is your story, let me just offer a few words of encouragement.

First, understand the landscape of your competition and pain points of your customers. You need to understand your industry inside and out. And you need to be able to anticipate your competitor's next moves. Remember that innovation always starts with the customer. Focus groups and surveys are fine, but they never go far enough. Instead, it's imperative that you go to customers and see how their products or services are used. Discover their unique pain points, and design your innovations around reducing or eliminating these issues.

Second, remain focused on a clear and compelling vision. Whenever I talk about vision, I think some people view this as being able to see twenty years into the future. But that's not necessarily the case. Often, vision is just being able to see around the corner. And as you develop your vision, you must communicate it effectively to those you lead. It should frame everything you do. If it doesn't, those on your team will assume your new innovative idea is just the flavor of the month. For critical projects, pick the best people and ensure they are 100 percent dedicated to the project. Along with this, place people from all the functional groups on the project. If this innovation is your ticket to growth, then it must be understood by every group and flow smoothly from development to implementation.

Third, live with passion. Your vision should excite you, and your excitement should bubble over to your team. When you speak, others in and outside your organization should sense your optimism. Passion flows from how you talk, ask questions, and the body language you display.

Fourth, be resilient. Yes, there will be times the giant of innovation feels like it's about to crush you. But keep going. Be like that toy clown that keeps popping back up. Don't be afraid to seek outside advice.

And fifth, take time to celebrate. Years ago, I sat in a room as various leaders shared their end-of-the-year reports. And nearly every person had positive numbers to share. Despite this fact, the boss stood up at the end of the meeting and gave a short speech that went something like this: "Good year everyone, but now here is what you did wrong." A few minutes later, I rose to my feet and said, "Guys, we've just had one of the best years in our company's history. Can't we all just have a moment of joy?" And since that moment, that phrase has always stuck in my head. Life is hard, business is hard, and innovating is hard. But amid turning giants, take time to celebrate your wins.

Innovation is a culture. It requires consistent courage. And it's a never-ending process. Don't be afraid to seek advice from the outside. Because just as you start to win, a new giant comes along.

CHAPTER 4:

THE GIANT OF SKEPTICISM

The giant of skepticism isn't exactly new. It's been part of our human social fabric from the beginning. As people, we love to question, discuss, and debate.

And as a result, nearly every breakthrough discovery was met with skepticism. Two thousand years before Columbus sailed the ocean blue, ancient Greet philosophers like Pythagoras and Eratosthenes were ridiculed for concluding the Earth was not flat. In 1543, Nicolaus Copernicus first presented his theory that Earth rotated around the sun. But this breakthrough belief took more than a century to find widespread acceptance. When French chemist and microbiologist Louis Pasteur postulated microorganisms were the cause of certain diseases, many thought this suggestion was ludicrous.

Consider modern innovators like Elon Musk. Regardless of your personal views, it's amazing to see what people like him have accomplished. Despite intense pushback and seemingly overwhelming odds, Musk has accomplished so much. Just a few miles from where I live in

Austin, he constructed a Tesla giga manufacturing plant that I believe is the largest factory I've ever seen. And if I take a six-hour drive south to Boca Chica, Texas, I'm able to see another one of Musk's innovative programs, SpaceX. And even through his 2022 acquisition of Twitter, it's clear Musk has learned to turn the giant of skepticism. And this brings us to an important point.

The twin sibling of innovation is skepticism. From the invention of the airplane, mass production of automobiles, and advancements in modern technology, there will always be a certain segment of the population resistant to change. *Won't this cause more harm than good? Who is going to pay for it? How do we know it will ever work?*

My entire life I've had skeptics. In the early years of my career, when I was changing jobs every three years, my parents were skeptical. Growing up during the Great Depression, they felt I should consider myself lucky to have *any* job and that I should just stick with whatever path I started.

If you're an innovator, it's easy to fear skepticism and let it overwhelm you. But as I've noted, when faced with a task that feels insurmountable, it's important to break it down into bite-sized chunks. One of the ways to do this is by identifying the different types of skepticism you face.

FOUR LEVELS OF SKEPTICISM

We've already addressed the internal giant of skepticism, or doubt, in chapter 3. This one is the hardest to overcome. But when you're confident in yourself and committed to persistence, the next level of skepticism includes those external forces. It's the voices around you doubting your ideas or if you have what it takes.

When I launched the DeltaV project, I had skepticism in spades. And there were four primary types of skepticism I faced.

TYPE 1 | SKEPTICAL MANAGEMENT

Going back to my conversation with Chuck Knight, it's not as if I quelled all his doubts in our first conversation. Far from it. In fact, his exact words were, "John, if I don't understand this, it's your fault."

I took his challenge to heart. His words made me refine my pitch and focus on the two most important elements—helping customers and differentiating from the competition. I couldn't just speak in technical terms. I had to speak like a businessman if I was going to get others to buy in.

And buy-in was often tough. A few organizational leaders thought of me as nothing more than a propeller head—someone too absorbed in technology and not grounded in reality. And to be fair, this was a common problem among engineers. Engineers often fell in love with their own innovations and thought, *I've got a solution. Now I just need a problem!* That's how some in management saw me. I was "Berra the propeller head"—a nerd who was too clever for his own good.

When word of our idea got around, I got wind of what those in other parts of Emerson were saying. Some were positive, some were skeptical, and others were a bit nasty. One of the common pushbacks was that I was trying to push a technology customers weren't ready to embrace. I'll never forget a remark from one gentleman I respected who told Chuck, "John is a marketer and salesman. He's pitching you an idea that is never going to happen. It's just a bunch of hooey."

A bunch of hooey.

The funny thing was the gentleman who made this comment was someone I considered a friend. I'd even helped promote him.

And hearing statements like these often made me pause and scratch my head.

Some of the skepticism was easy to explain. Our workplace was competitive. And leaders of other departments within Emerson likely feared what might happen if my ideas took off. They were insecure or jealous. And there were times it would have been easy for me to take the bait and punch back with a sarcastic comment or two of my own. But whenever possible, if I heard someone make a negative comment, I'd pull the person aside and try to explain what we were trying to do. Often this made a big difference.

I didn't magically turn the giant of skeptical management in a day or even a year. Rather, it was a series of small turns in one-on-one conversations that made the difference. Persistence was the key. As I continued to share the vision, continued to deliver on promises, and continued to help our existing business thrive, this slowly turned the tide of skepticism.

A final word about skeptical management. Although it was tough to get my ideas through to leaders like Chuck Knight at Emerson, I did get them through. However, the culture of Emerson at that time was not for the faint of heart when it came to innovation. Presentations and division reviews were conducted as a contact sport. And a culture like this kills many ideas, because people are too afraid to speak up.

So, this raises an important question: If you are in a management position, what sort of culture are you leading?

TYPE 2 | SKEPTICAL EMPLOYEES

When I first shared this DeltaV vision with the employees at Fisher-Rosemount Systems, not everyone was thrilled. There were people in key positions who thought it was next to impossible to go from where we stood to the picture I had painted. In the back of their minds, I think many thought I was delusional. And if my idea didn't pan out and they supported me, they too would be caught up in my failure.

As a result, I had employees start to hold back. They were a bit like members of Congress who vote "present" on a bill. They didn't *oppose* my new plan, nor did they *support* it. They were there in body but were not going to lend their full efforts and reputation to something they did not fully believe in.

Also, what made matters especially tricky was that I ended up selecting around fifty of our best people to work on DeltaV. This team worked off-site, had relaxed dress codes, and worked on our grand futuristic project. The rest of the employees remained at our facility in Austin and focused on maintaining our current products and helping our $200 million organization function. Naturally, some of the employees who weren't selected to join the new team felt resentful.

On top of this, it was clear many of the employees still felt beaten up by previous failures. For years, Fisher and Rosemount had done all they could to grow in the control systems market. But progress was slow, and they couldn't make a dent in incumbents. They'd crafted countless proposals and offered hundreds of demonstrations to potential customers, only to have most of those customers decide to do business elsewhere. As a result, many Fisher-Rosemount Systems employees were skeptical about trying something new.

Knowing everyone was in shock and that I needed to earn their trust, I made it my top priority to meet one-on-one with key leaders. I asked the staff at Austin and Minneapolis to make me a list of the top ten people who had shown a lot of potential. Soon I had a list of about fifty names, and I set up an hour-long meeting with each of them. During the first few minutes, I'd say, "Listen, I hear you've done great things here, and I want you to know our company values you. My door is open if you ever want to talk about anything." As soon as

I said this, I'd see this look of relief wash over their faces. They knew I wasn't there to put them out to pasture.

Second, I looked at every individual and said, "You've been working here a long time. What advice would you give *me*?" For some, they weren't used to hearing a question like this from a boss. But I quickly generated a long list of remarkable ideas to improve our company. And almost more important, this simple question opened the door for many to address their fears and concerns. Through these meetings, I learned a lot about the tone and culture of the people I was asked to lead. For example, upon hearing this question, one of the men looked at me and said, "The best advice I can give you is to pick a strategy and stick with it. I've been here six years, and you're the third person in this job." That comment stuck, and I resolved to bring a sense of stability to this new culture.

Next, I started sending out a monthly memo to our employees. Nothing long, but there was enough content to keep everyone informed and on mission. I'd talk about what our strategy was going to be now that we were a larger organization. And I would address concerns I'd heard employees share.

I also conducted several all-employee meetings in both locations. As it turned out, this form of communication proved very helpful. Keep in mind this was before the days of email and social media. Employees couldn't check my Instagram or Twitter feed to know if I was out of town. And whenever I left Austin, people in the company wanted to know where I was and what I was doing.

Later on, I jumped on the technology of webcasting. I'd do the Americas, Europe, and the Middle East in the morning and repeat in the evening for Asia Pacific. We used a chat feature to let attendees send questions while I was presenting, and then I would answer them at the end. Opening the doors of communication helped quell many of their fears.

In *Leaders Eat Last*, Simon Sinek writes, "You can easily judge the character of a man by how he treats those who can do nothing for him."[17] It would have been easy to step into Fisher-Rosemount Systems and lead with a heavy hand. After all, changes needed to be made, and there was only so much time to ensure every person sat at their right seat on the bus.[18] However, I knew that wasn't the approach I wanted to take.

For one, that's just not who I am. And so I made an intentional effort to avoid eating lunch in my office and instead walked down to the cafeteria and sat with random groups of employees. It was there I learned people's names, what they liked, and how they lived. I asked them what they were working on and how long they had been at the company. Because I was new to the area, I spoke with them about what it was like to live in Austin. The reality was there were times I had to make tough decisions and was forced to keep one person over another. But amid this upheaval, I did my best to project a sense of calm and treat employees the right way.

The next step was training and team building. At Rosemount, we'd used a Spencer, Shenk, and Capers personality course to assess ourselves and those on our team. As someone who had a personal relationship with Mr. Capers, I found him unusual in his ability to pinpoint a person's strengths and weaknesses. Mr. Capers was a psychologist, and I had him walk our team through several days of extensive training using a course called Managing Motivation for Performance Improvement. And when his sessions were over, I had a much better read on key leaders in our new organization—including myself.

---

17   Simon Sinek, *Leaders Eat Last* (New York: Portfolio, 2014).

18   Jim Collins, *Good to Great* (New York, NY: HarperCollins Publishers, 2005).

Next, I set up an off-site meeting at a hotel. We had a brainstorming session over the course of a few days and stood several easel pads along the edges of the room. I told everyone that there were no bad ideas and that I wanted to hear their thoughts. *What was Fisher-Rosemount Systems doing right today? What were we doing wrong? What were we not doing that we should be doing?*

After several hours we'd collected hundreds of responses. And while we obviously couldn't implement every one of them, we asked each person to pick their top five. From here we distilled these down to some excellent action items, which helped all of us get on the same page and improve the performance of the company.

And step by step, I gained the trust of those in this new organization. As team members felt empowered to step into their new roles and understood clear expectations, the cloud of skepticism that hovered over Fisher-Rosemount Systems subsided.

TYPE 3 | SKEPTICAL CUSTOMERS

This third type of skepticism came from the thousands of clients we served around the globe. Keep in mind most of them were still getting used to Fisher-Rosemount Systems as a joint operation. And we heard no shortage of remarks about our infamous blue hyphen that joined these names together.

And even though our combined system companies had a comparatively low share of the market, we certainly cared about what our customers thought and did not want to alienate our base. One of the common fears many had when we merged was whether they would continue to receive the same level of service as before. If X company in Cleveland had purchased a product from Fisher, *would the merger of Fisher-Rosemount Systems mean their system still worked the same? Or would it falter? Would we decide to go with Rosemount's System and drop Fisher's?*

Customers who had purchased Rosemount Systems had the same questions. Both sets of customers who had bought either Fisher's or Rosemount's control system were now concerned we would opt for the other control system and leave them high and dry.

To alleviate our customers' fears, we developed a long-term support plan for both systems. And not only did this ease our customers' skepticism, but it also gave us added credibility as we anticipated the launch of DeltaV. Like most organizations, Fisher-Rosemount Systems had three types of customers: those who *embraced* innovation, those who were *suspicious* of innovation, and those who *opposed* innovation.

Understanding this reality, I turned to our Fisher-Rosemount Systems' salesforce and asked them to identify a list of early adopter potential customers. These would be the customers who were excited about something new and give our DeltaV system an honest review. From there we developed a customer advisory panel to get early and direct feedback on our efforts and to serve as beta testers prior to launch.

Our goal was simple. By focusing first on customers who embraced our technology, we could collect some important feedback and troubleshoot many of the problems we encountered before taking this product to customers who were skeptical or opposed innovation.

Bill Gates' testimonial certainly didn't hurt. And the Gartner Group evaluated our project and gave us one of the highest ratings for vision. Along with this, I made sure we worked all magazines and trade publications. We did a ton of unconventional advertising, and I signed up for any roundtable discussions or industry forums that would get the word out about our innovative system. Basically, if anyone wanted to talk DeltaV, I or someone else from Fisher-Rosemount Systems was willing to fly out and participate.

John making one of his many talks to the industry

If we talked to a customer or someone who was interested in our system and they still had doubts, we'd invite them to visit Austin and try out our demo system. For example, I remember one customer who was worried about cybersecurity. Knowing these concerns, we invited him to bring along a CD or memory stick that included any viruses they'd encountered with their current systems and see what happened. As it turned out, DeltaV passed with flying colors.

Prone as I was to overcomplicating a presentation to management, I knew I couldn't make the same mistake with our customers. And thus our sales pitch was "It's really easy" with a picture of a baby using one of our devices.

Today, that picture looks cheesy, but keep in mind this was the mid-1990s, with Windows 95 yet to be released. Chuck Knight's assistant was still typing out company memos on an IBM Electric typewriter and sending them to key leaders on a fax machine.

But as the internet grew more pervasive, we came up with an idea to send customers to a website. On this site, we built an entire plant simulator application. At the time, there was a popular video game called *Sim City*, and so we piggybacked off that concept. When customers arrived at our site, they were asked to enter some key data about their plants, including the processing equipment they had and the physical size. The website would then compute how much money they would save if they went with our complete offering, including our system, measurement products, and control valve products. Our innovations saved money on wiring costs, control room size, and increased uptime through the use of diagnostics to predict issues before they happened. Was it perfectly accurate? No. But it did get the conversation started.

Granted, while this option was innovative, fun, and informative, it was a tough sell. Customers were skeptical about the calculated savings. To combat this skepticism, we even offered customers a guarantee that, in retrospect, was a little risky. We said we would promise the results if the customer would let us baseline their plant performance that day. We offered this knowing few companies would allow us to do a baseline on their plants. Most of them already knew there was room for improvement through better automation, but they had always postponed these investments. There was also the issue of them potentially exposing proprietary information.

No one ever took us up on this guarantee, but this level of confidence we displayed assured companies we were serious.

TYPE 4 | SKEPTICAL COMPETITORS

If there was one kind of skepticism I didn't mind, it was the kind we received from our competitors. None of them, especially Honeywell, took us seriously. Many referred to the DeltaV project as a toy or science experiment.

But the fact they ignored us worked in our favor. To me, we were building the Great Wall of China. People could laugh and mock us. And they could assure themselves the "toy" would never result in a real control system. But I didn't care because I knew with enough effort and enough time, we would have the last laugh.

We were creating our own innovator's dilemma. As we grew, if competitors tried to copy us, it would pose a major problem for their installed base. In the systems business, the installed base is like the razor, and upgrades and parts are like the blades. A supplier could generally count on 7–10 percent of the installed value of the system every year in parts and small upgrades. And the prices on these annual expenditures were higher, sort of like what you pay for parts on your car. This meant copying DeltaV would be costly because they would not only have to spend money on development, but they would be jeopardizing some of their annual parts and upgrade revenue.

When you look at some of our major competitors such as Honeywell, Bailey, Taylor, and Foxboro, we only had installed bases that were a fraction of the size. Some of our competitors had five times the number of customers we served. On the surface, this appeared a disadvantage for us. But when it came to innovation, I knew this gave us the edge because we could afford to take risks other organizations couldn't.

Part of our competitive advantage was that we started slow and didn't present ourselves as a major threat. We created what we might call a "baby DeltaV." On the surface, it didn't appear that intimidating. But I knew that while we might start small and only work with a few applications, it was only a matter of time before we grew.

We'll talk about this more in the next chapter.

HEALTHY AND UNHEALTHY SKEPTICISM

There is healthy and unhealthy skepticism. "Healthy skepticism," David Allan writes, "is often the best way to glean the value of what's being presented—challenge it; prove it wrong, if you can. That creates engagement, which is the key to understanding."[19]

Skepticism isn't wrong. It sharpens the innovator. Skepticism rids you of poor hobbies and habits. Show me a business that hasn't faced any skepticism, and I'll show you a business that isn't running at its optimal level.

---

19   David Allen, *Getting Things Done: The Art of Stress-Free Productivity* (New York: Penguin Books, 2001). Kindle Location 350.

Think about the last time you visited a website that felt like it was created twenty years ago. The user experience was clunky. Whenever you clicked one tab, you lost your place and couldn't easily get back to where you were. General information that should have been readily available felt like it was tucked away where no average person could find it.

Whenever I have this experience, I often think to myself, *Doesn't this business care about their customers? Don't they have one person who will tell them the truth?* And that simple poor experience leads me to doubt the credibility of the brand represented.

This is why every organization needs an appropriate amount of skepticism. Take, for example, our first iteration of DeltaV. It was far from ready for prime time. In fact, when we had our first customer advisory panel give it a go, they had a lot of challenges. But thanks to their feedback, our actual launch lived up to its promise. Healthy skepticism caused us to rethink how our system was designed and allowed us to make some significant changes before we unveiled it to the masses.

That said, there is a type of skepticism that is just wrong. It's bred out of jealousy and inferiority. And certainly, I faced my share of these comments. "Did you hear what Berra wants to do? He wants to put an operator station in a PC!"

But in my mind, I wasn't confronting anything innovators of the past hadn't faced ten times over. I think back to great Serbian American inventor Nikola Tesla and his concept of alternating current, which stood as a rival to Thomas Edison's direct current. Edison went to great lengths to show how DC was superior to AC, even arranging for a convicted criminal to be killed in an electric chair using AC power. The message to the public was clear: put these wires in your homes and this is what can happen to you.

Still, at the end of the day, Tesla's model is the one that won the day and now resides in nearly every American home. It just took many years of turning the tide of overwhelming skepticism. This is the same challenge every innovator faces today. Skepticism is par for the course. So accept and embrace it.

That said, a good litmus test for evaluating skepticism is to consider the source. For example, when Chuck Knight told me I wasn't making myself clear in my presentation to the Emerson leaders, I used his comments as fuel to get better. But when jealous individuals made back-stabbing comments designed to throw me off my game, I let these pass.

In most cases, healthy skepticism comes from healthy people. Unhealthy skepticism comes from unhealthy people. The key is to surround yourself with the right voices and take their feedback to heart.

TURNING SKEPTICISM

Clearly, the first and foremost way to overcome skeptics is to persist and use their skepticism for fuel. It's having an *I'm going to prove them wrong* mentality. And from here, there are several keys to turning the giant of skepticism.

First, accept that any innovation you come up with will have skeptics. If it doesn't, it's probably not innovative. You're going to face skepticism from management, peers, employees, and competition. This is just a reality. Embrace it.

Second, listen to your skeptics' greatest concerns before you try to turn them. It's important to understand and accept the most valid criticisms skeptics have to offer. This goes back to the importance of working with early adopters and allowing skeptics to test your idea before it's presented to the masses. Don't expect others to place blind belief in your innovations. Deal one-on-one with skeptics rather than launching into your pitch. Ask yourself: *Why are they skeptical?*

Third, systematically turn your skeptics one conversation at a time. Sometimes skeptics' doubts are personal. They might say something like, "I don't think you have the leadership qualities to make this happen." Statements like this hit home, and if we're not careful, we can take them too personally and offer a poor response. Instead, this is where it's important to have additional conversations. Know what questions your skeptics have, and then systematically address their concerns.

Fourth, develop a customer advisory panel. Find people who are on the front lines of your industry, and find creative ways to collect their feedback on your product. Show them steps along the way, and have them comment on your innovative ideas. Let them try your prototypes. Make them part of your extended team. And before you launch, have them offer testimonials of their experiences to quell the skepticism other customers might have.

Fifth, keep getting back up. Remember the example I shared of the inflatable toy clown that keeps getting back up? That is how you need to handle the punches of skepticism. Some of them hit hard, but the key is to get back on your feet and keep moving toward your vision.

BE A TIMELY VOICE IN SOMEONE'S LIFE

Remember what I said at the end of the last chapter about the importance of taking time to have "moments of joy"? I want to build on that and remind you to take time to celebrate the successes of others, even if they don't report to you or aren't in your organization.

Looking back on my life, two examples come to mind. One positive, and the other negative. The first occurred in grade school with my friend Mike. At the time, there was a college prep high school called St. Louis University High. It was an all-boys school and was considered the best private high school in the city. Several guys in my class were interested in trying out for it.

I wasn't. In my mind, I didn't stand a chance to get in. And if I did, my parents certainly wouldn't be able to afford it. But one day, my friend Mike walked up to me and said, "John, you need to take the test. What have you got to lose? If you get in, it's another option to consider. But if you don't apply, you never know what could have been."

Based on this short conversation, I applied and took the exam. To my surprise, I was accepted. And for the next four years, my parents made enormous sacrifices to cover my tuition—even borrowing against their whole life insurance policy. While I couldn't see it at the time, that word of affirmation from Mike and the willingness of my parents to sacrifice made my career possible. Without those years of education, my entire life would have looked very different.

Contrast this story with one that happened when I was in college. I was taking an engineering design class and was assigned to develop something that would work in the marketplace. At the time, I worked at a specialty retail store that sold Italian meats and cheeses. Customers would come by and make their orders for sliced salami or prosciutto. As the worker behind the counter, I was tasked with slicing the food, putting it on a scale, cross-referencing the weight with the price per pound, wrapping up the package, and writing the price on the package.

After a few experiments, I came up with an idea to create a device that combined a scale that used electronic technology with a calculator and a printer. All the butcher had to do was enter the price on the keypad and the scale would print a label with the purchase price on it.

In the end, I got a C- on the project, and my professor said the idea was "too far-fetched." I soon forgot about this idea. Of course, today this technology is common. In fact, next time you are at a grocery store or deli, look at the integrated scale that is used. Sometimes I look back and wonder what could have been. What would have happened if my professor had encouraged rather than squelched my idea?

Maybe you're a skeptical person by nature. That can be a good thing. But learn to temper your skepticism in healthy ways. Speak to build up rather than tear down. Be the timely word others need.

CHAPTER 5:

THE GIANT OF COMPETITION

Most people see the giant of competition as a bad thing. But I don't think of it that way. When I consider the giant of competition, I see it as a tool to make me sharper. It's a powerful motivator that increases my focus. In the words of humorist and businessman Arnold Glasow, "Without the spur of competition we'd loaf our life." Competition makes us better because it urges us to reach new levels.

Competition even predates skepticism. Turn on National Geographic, and almost any animal species you see on the screen will display some level of competition. Male creatures from gorillas to kangaroos all spar with one another to achieve dominance and claim rights to their female counterparts. Competition is everywhere.

Business competition is unique in that there is an important third party—the customer. Business competition is all about making your company or organization look more attractive to consumers. The customer decides the winner, and everything you do that touches the customer is a competitive move. While competition in business is more complicated, it is still intense.

Every thriving business in the developed world faces some threat of competition. In his seminal work on competition, Michael Porter said that all business competition can be placed into one of three categories—cost, focus, or differentiation.[20]

But these three categories Porter listed in 1985 don't cover the complexities of today's competitive landscape. Competition today is more like competitive chess. Perhaps you've watched a video clip of a grandmaster chess champion like Magnus Carlsen playing ten games at once. It's astounding to watch as he starts with one board and then moves to the next, seamlessly keeping track of the progression of each game.

Business competition is like this. It's playing multiple games at once, measuring the effectiveness of each move, and keeping track of opponents' countermoves. But what makes business competition even more challenging is that each game interacts with the others. Everything is interconnected. A move made on one board has ramifications for another.

No one is a master of competition. As Porter notes, it is hard to be excellent in *every* category. Competitors are fluid. They get better in some areas, and then new competitors emerge to challenge the incumbents.

---

20   Michael Porter, *Competitive Advantage: Creating and Sustaining Superior Performance* (New York: Free Press, 1985), 11–15.

BE PARANOID INSTEAD OF COMPLACENT

There are two extreme responses to competition: complacency and paranoia. I always leaned toward the latter option and subscribed to the title of Intel CEO and Chairman Andy Grove's book that *Only the Paranoid Survive.*[21]

My paranoia served me well and drove me to understand what our competitors were doing. I wanted to know their track records, their level of investment in new technologies, and how their customers viewed them. The bad aspect of this paranoia meant I had a lot of sleepless nights, but the good by-product was intense focus. Every time I asked someone on our team a question, I did so with our competition in mind.

Competition defined the way we established our goals at Fisher-Rosemount. We never measured our progress against what we'd achieved in the past. Rather, we measured our progress based on where our innovations placed us in the marketplace. To me, developing a product that was twice as good as our previous generation meant nothing if we were still miles behind our competitors.

Failing to take note of the competition is a strong indicator of complacency. And complacency is a fast track to failure. There are so many examples I could include here. A few that come to mind include the dramatic collapse of Lehman Brothers, the loss of market share by Nokia in the mobile phone market, and the loss in the consumer film industry by Kodak.

A common theme with each of these three examples is complacency. It's sort of that whole, "too big to fail" mentality.

---

21 Andrew Grove, *Only the Paranoid Survive: How to Exploit the Crisis Points That Challenge Every Company* (New York: Currency Doubleday, 1996).

And to be transparent, there have been times I was complacent. I think back to my days before the Fisher-Rosemount merger when I worked at Rosemount. During the 1980s, most of the team I worked with thought microprocessors worked great in a tame environment, such as a control room. But everyone was skeptical they would operate correctly out in the elements.

In retrospect, this feels so foolish because almost every vehicle today has dozens of microprocessors under the hood, and they function just fine regardless of whether the car is in Alaska or Texas. But at the time, I just rode the tide of popular belief. And it wasn't until Honeywell introduced field devices with microprocessors that I changed my perspective. By then, we were far behind the innovation curve and ended up paying a price for our complacency and lack of paranoia.

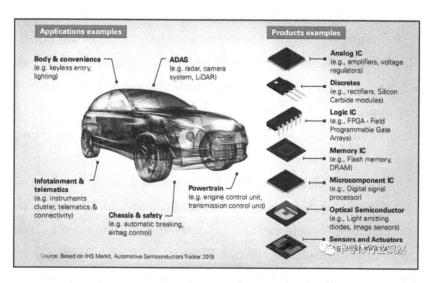

Chips in a car

DON'T BECOME ARROGANT

Leaders who are too paranoid become dictators. But leaders who aren't a bit paranoid become arrogant. They underestimate their competitors and think they have all the answers. History is full of such examples.

England's arrogance said General Washington and his inferior army were no match for the likes of the highly trained British infantry. Hitler's arrogance to stop trusting his generals led him to think he could fight and win a war on two fronts. And in recent years, almost every major military miscalculation has had some level of arrogance attached.

It's a short distance from competence to arrogance. Some leaders start great, have wonderful years, and then fall prey to the corruption of absolute power. They lose their touch, and as a result, they lose touch.

Some leaders work well when they're managing up. However, when they're elevated to a higher position, they become more dictatorial. They knew how to be charming and do all the right things to get the job. But now that they have it, they micromanage and control others. These leaders tend to remove people who question them, whereas a real leader should do just the opposite.

The key to not becoming arrogant is to cultivate a few key relationships with people who report to you. And these relationships need to be such that they can come to you one-on-one and talk about any mistakes you make. These individuals are invaluable because they help you understand how others perceive you. When your head of sales walks into your office and tells you that you don't know your butt from a hole in the ground, it keeps you humble.

Personally, I had two people like that who did not hesitate to tell me what I needed to hear. And similarly, I tried to be a positive voice for those above me, such as the CEO of Emerson. I never criticized my superiors in public, but I did my best to go in and explain when I had a difference in opinion.

A primary example of business arrogance can be found in the history of IBM. There was a time IBM was at the pinnacle of the mainframe computer world. But they grew complacent and didn't pay enough attention to their emerging competition. They underestimated the impact of personal computers. And IBM subcontracted their PC operating system to Microsoft, which enabled Microsoft to license its PC operating system to IBM competitors.

But in 1993, the same time I was adjusting to my new role as president of Fisher-Rosemount Systems, Louis Gerstner became CEO of IBM and made sweeping changes that saved the company. He ordered thirty-five thousand layoffs, reduced overhead costs, and transformed IBM's culture. He saw that customers were looking for solutions and organized the company around industry verticals, such as banking or manufacturing.

A large part of shifting this culture was a renewed focus on execution. In Gerstner's words, IBM's turnaround was "all about execution."[22] In July 1993, Gerstner identified four major strategies that would prove crucial to IBM's turnaround.

1. Keep the company together—This strategy was implemented in order to help IBM to utilize its competitive advantage (resulting from its scale) and offer integrated solutions to clients.

2. Change the fundamental economic model—This strategy started by comparing expense-to-revenue of IBM with its competitors. Because of higher expenses at IBM as compared to competitors, a massive program for expense reduction was launched.

---

22 "Case Study: IBM's Turnaround Under Lou Gerstner," MBA Knol, accessed August 4, 2023, www.mbaknol.com/management-case-studies/case-study-ibms-turnaround-under-lou-gerstner/.

3. Re-engineer how business was done—Gerstner saw that the IBM processes were cumbersome, highly expensive, and redundant. He introduced a re-engineering initiative which drastically reduced the overhead expenses of IBM.

4. Sell underproductive assets in order to raise cash—Under this strategic objective, IBM sold off unproductive assets to raise cash.[23]

While these strategies were the core behind Gerstner's success, it wasn't until later that Gerstner identified the real reason IBM changed. In *Who Says Elephants Can't Dance?* Gerstner makes this admission:

> Until I came to IBM, I probably would have told you that culture was just one among several important elements in any organization's makeup and success—along with vision, strategy, marketing, financials, and the like ... I came to see, in my time at IBM, that culture isn't just one aspect of the game, it is the game. In the end, an organization is nothing more than the collective capacity of its people to create value.[24]

And by transforming IBM's arrogant culture, Gerstner repositioned IBM to effectively compete in the marketplace.

DEVELOP YOUR *COMPETITIVE* *ADVANTAGES BIBLE*

If you're wondering how to compete effectively, I urge you to develop what I call a *Competitive Advantages Bible.*

---

23   Ibid.

24   Louis V. Gerstner, *Who Says Elephants Can't Dance: Leading a Great Enterprise through Dramatic Change* (New York: HarperCollins, 2002).

Before we ever wrote our first line of code for DeltaV, we developed a document that was around twenty-five pages long and defined the competitive advantages we wanted to have over the existing control systems competition. This included all the pain points our customers had and how our system would solve these challenges.

We were committed to DeltaV being a new system and not merely an evolution of an existing one. We knew that to truly disrupt the industry, we had to start with a clean sheet of paper. And our view was if we weren't disrupting the industry, someone in the industry would disrupt us.

When we had everything in writing, this crystalized our thought process and goals. We were able to evaluate in real time whether a potential opportunity was a good fit. Because we knew where our competition stood and how we stacked up against them, we operated from a position of strength.

As I've often said, the key to turning the giant of competition is to understand the competitive advantages you have and strive to build on them and create new ones. In the words of legendary CEO Jack Welch, "If you don't have a competitive advantage, don't compete."

That said, there are three keys to turning the giant of competition.

KEY #1 | DEVELOP A STRONG AND
PERPETUAL COMPETITIVE ANALYSIS

As my role at Emerson evolved, we tried to make competitive analysis an art form. This meant we had people dedicated to writing and maintaining an analysis of each of our main competitors. These individuals gleaned everything they could from public sources. They visited competitors' trade show booths and took detailed notes about what they saw and heard.

My executive staff joined trade associations so that we could see competitive executives face-to-face and try to gauge their personalities. We accepted invitations to participate in industry forums and roundtables. And often these discussions provided valuable information. The Q&A sessions in these events provided unique insights into the priorities and personalities of our competitors. Of course, we also realized we needed to be mindful of what we said, knowing our competitors were listening to us as well.

Because companies take on the values and beliefs of their leadership, we did our best to draw each competitor's organization chart. And if the competitor was part of a larger organization, we'd assess how much love that business got within the corporation. We also talked to customers, another great source of information. We did not ask them to reveal confidential information, but we listened to what they liked and disliked about other brands.

We did a gap analysis that compared ourselves to the competition and assessed where we fell short. This, in turn, generated a list of possible actions or acquisitions to fill that opening.

As I see it, a business needs to do the same deep dive into its competition as it does its customers. To this day, it frustrates me when I sit on the board of some organization and the entire presentation revolves around "new and improved" concepts. Everything talked about has to do with comparing their latest and greatest products with previous generations the company produced the year before. There is almost no discussion of what competitors are doing and how this latest innovation stacks up against them.

Today, whenever I sit in a presentation like this, I always ask the presenter to list the "silver bullets" or provide an "elevator speech," which describes how they stack up against current and anticipated competition. If they can't, I know they're in trouble.

It's imperative to avoid this mistake. Don't compare yourself to yourself. Compare yourself to your competition. And the only way you can do this is through a rigorous, repetitive, competitive analysis.

KEY #2 | DIFFERENTIATE YOURSELF IN AS MANY WAYS AS YOU CAN

Think broadly about how you compete. Most competitive analyses I see address the product or service. But you need to go beyond your product or service and find ways to differentiate yourself from the competition.

There are numerous ways to do this. Businesses today compete in so many different areas. They compete on:

- Talent
- Financing and capital structure
- Product or service
- Cost of goods sold
- Supply chain
- Marketing and promotion
- ESG
- Social media and website
- Ease of doing business
- Selling organization
- Customer service
- Customer-facing software
- Quality
- Brand
- Reputation
- Acquisitions

Even this list is not all-inclusive. These are the multiple chess boards that form your competitive game. If your product or service has a lower cost, you've got an advantage. If there is something you can do that your competitors can't easily replicate, you've got an edge. If you have award-winning customer service, this is a great way to set yourself apart from the pack.

There are countless ways to differentiate yourself.

Let me give you an example that has to do with talent. Eventually, when I ran the entire Emerson automation business (which I'll explain in the next chapter), I asked each divisional president to come to a specific meeting with a list of high potential people in their organization. Everyone was in the room together as we listened to each president describe their "hi-pots" and identify what sort of potential they had.

My vice president of human resources and I kept track of all the people on the list, and so did the presidents. And from there we encouraged and monitored each one of these individuals' career paths. If one president had a need and there was an ideal person in another division, we encouraged them to move. This was better than losing them to an outside company with no affiliation.

As this story illustrates, talent development is one of the primary ways to differentiate yourself from the competition. Unfortunately, I see so many CEOs who have failed to develop successors, and they end up paying a steep price in the end. Talented people differentiate your company in so many ways.

KEY #3 | *KNOW YOUR TARGET AUDIENCE*

Ray Kroc said, "Competition can try to steal my plans and copy my style. But they can't read my mind; so I'll leave them a mile and a half behind."

One of the best ways to gain a competitive advantage is to know your target audience inside and out. Sometimes I've failed miserably at this, and I think back to my time at Rosemount for a specific example. One day our Cleveland sales engineer assembled a large group of customers and asked me to give a demonstration of our control system. I was happy to do so and confidently launched into my pitch, telling my audience that since this system was configurable, it eliminated the need for expensive computer programmers.

There was only one problem. Half of the people in the room were computer programmers! If I had only known my target audience that day, I would have given an entirely different presentation.

This embarrassment taught me the importance of remaining focused on my customers' needs. If I would have understood whom I was speaking to, I would have shifted my focus to show how our system wouldn't have required the programmers to spend time on mundane code and allowed them to focus on important tasks.

ACTIVELY ENGAGE IN THE MARKET

Developing your *Competitive Advantages Bible* is important. Unfortunately, some leaders have an idealistic mindset and think they can hunker down and zero in on what their company is creating, without giving a second thought to the competition. They believe they're being focused and avoiding outside distractions. But this outlook can prove disastrous.

From my vantage point, there are several crucial ways to engage in your market.

FIRST | SHAPE YOUR INDUSTRY'S STANDARDS

First, if you consider yourself a player in your industry, you should be part of shaping the industry's standards.

Here is what I mean. Many industries have standards associations or groups. And in a world of industrial manufacturing and IT, standards play a huge part in determining what fits and what doesn't. A simple example of this is the differences found between the giant electrical plug in the United Kingdom, the continental pin plug in

Europe, and the prong plug used in the United States. Another is the United Kingdom's recent law that says all cell phone chargers must be USB-C. This means companies like Apple must adapt and switch from their lightening connectors or risk being phased out of the marketplace.

While many companies view the standardization process as a glorified Washington, DC bureaucracy that isn't worth their effort, I think it's imperative to pay attention and participate in shaping the industry. And here is why.

Standards-making bodies usually consist of representatives from end user and supplier companies. Although the representatives are supposed to have the best interests of the industry at heart, it is virtually impossible for them not to think of how the emerging standards will impact their companies and to push for standards from that perspective.

As mentioned in previous chapters, the emergence of digital communication from the field to the control room was a bumpy process. Some suppliers started out with proprietary schemes that could block competitors. Imagine a world where an Apple iPhone user could only talk, text, or email to other Apple users. This was how digital field communication started in the automation world.

At Emerson, we saw that customers didn't like this, and neither did we. This prompted us to drive open digital field communication by developing technology and donating it to foundations to manage. For example, I was chairman of the Fieldbus Foundation from its inception until I retired. And in this foundation we drew together both end users and suppliers and worked out a technology all could use. The foundation hired engineers to test compliance and register each device that passed the test. It was important to lead this process.

If there are emerging standards in your industry, I urge you to jump in and show leadership. Send your best people to the meetings, and keep a close eye on the development.

SECOND | *PARTICIPATE IN BROADER DISCUSSIONS*

Next, remain active in the broader discussion. Most industries have a forum and are covered by trade or regular press. My encouragement to you would be not to dismiss this as a waste of time. Participation in these events is part of building your company's brand, and you should send your best people to attend.

Today, in particular, any participation is likely to be archived on the web and passed around on social media. Urge your technical people to publish papers or your key leaders to write op-eds. This can have a very positive effect on your employees as well as your customers.

I remember one forum where I was able to deliver a delightful zinger to the competition. The panel consisted of newly appointed leaders from the main automation companies, and they opened the session for questions from the crowd. An audience member approached the mic and asked the following question: "What makes each of you think you will be here next year?"

When it was my turn to answer, I responded a bit tongue in cheek. Looking at my fellow peers on the stage, I turned to the questioner and said, "Well, I'm the only one on this panel who was here last year, so I think my odds are pretty good."

THIRD | *ENGAGE ON SOCIAL MEDIA*

I know a lot of people in my generation who don't treat social media as seriously as they should. They view it as a place for cat videos and selfies. But social media is something that can help your competitive advantage. Like it or not, your customers use social media, and the use of blogs or podcasts will extend your reach.

Today, so many companies go above and beyond to connect with their customers. In addition to providing tech support, many offer online message boards where customers can help other customers troubleshoot issues. Not only is this a cost-effective strategy, but it also helps build numerous points of connection with those purchasing and using the products or services.

FOURTH | *VALUE YOUR BRAND*

A large part of establishing your brand and differentiating yourself from the competition is developing a culture of trust with your customers. For example, I have certain brands I'm loyal to and wouldn't think of purchasing from someone else. When I walk in a grocery store, I buy Heinz ketchup. Other brands might be just as good, but I've been loyal to Heinz for over forty years, and I don't think that loyalty will change anytime soon.

And then there are other experiences that make me pause and reconsider my loyalty. For example, I have a domain and an email that is hosted by a certain service provider. I've had it for over a decade, and it worked perfectly. But recently, I was informed that the provider was moving its hosting software from Microsoft Exchange 2013 to Exchange 2019. But the transfer process was clunky, and nothing seemed to work.

After hours on the phone and in chat rooms, I got a sheepish communication from the provider, saying they forgot to put something called an "auto-discover" into my domain name system (DNS). After they did, everything worked. But this process still made me scratch my head. *Was there no trial run or testing on this migration? How could they treat customers this way and expect to remain in business?*

Having a strong brand is important. And sometimes this isn't always easy, especially if customers confuse your name with someone else. For example, one of the challenges Emerson had was there was a similar company that went by an almost identical name. But the problem was this other company produced cheap consumer electronics. As a result, potential customers often confused us with them and had questions about our quality.

As I've shared, when I first took over as president of Fisher-Rosemount (the full automation group), our branding wasn't great. Customers were confused about what we offered and whether they'd receive the same quality of service they had in the past. At first, we felt it was important to hold on to the names Fisher and Rosemount because both had established market credibility. But eventually, we bit the bullet and called the company Emerson Process Management (now called Emerson Automation Solutions).

With each decision we made, we did so with our brand in mind. When you value your brand, you value your customers.

TREAT YOUR CUSTOMERS WELL

This brings us to another important point.

Many companies sponsor a customer event where the customers do the presenting. These are usually a few days long and include a couple of social events in the evenings. This exchange of ideas helps everyone because customers want to hear from other customers.

While COVID messed these sorts of in-person events up, they have since resumed and are highly effective. If you've never done one, start small and go from there. Along with this, it's important to think from the perspective of your customers. Try to place yourself in their shoes, and anticipate the questions or concerns they might have.

Unfortunately, sometimes companies do a horrible job at this. For example, a few days ago, my wife was on the phone with a certain bank that had been acquired by a larger bank. And as I sat in my office and thought through this section of the book, I could hear her frustration in the other room.

The transition was far from seamless. My wife had to reset her username, and her password had to be a certain length and include all these obscure characters. At one point, I heard her mutter, "Maybe I should just close this account and move it to another bank."

And as I listened to her frustration rise, I thought to myself that if this bank had really valued their brand, they would have put more effort into thinking through this transition. But it was evident that customer satisfaction was not as high on their priority scale as it should have been.

As president of Fisher-Rosemount, there were times our company messed up a major order for a client. And when this happened, sometimes my assistant or salesperson said, "John, we really messed this up and need you to talk to this customer." And so I'd get on the phone. Initially, some customers were angry. And many times, I realized they had a right to be upset. But taking time to listen was key. And usually after a few minutes, we were able to work out a solution.

If you are in a leadership position, always budget time to visit with your customers. It is a great way to learn how things are really going, areas for improvement, and plans for future purchases. It is also part of relationship building. Customer trust in your company starts with customer trust in you.

There was one customer visit I had many years ago that became a lasting memory. This was a time when the Soviet Union was breaking up and our European sales leadership thought it would be a great idea if the president of our company went to some "out of the way" places to show how important these potential customers were to us. One of these out of the way places was Tashkent, Uzbekistan.

Customer meeting in Uzbekistan, complete with gift clothing.
John is second from left, front row

At that time, Uzbekistan was part of the Soviet Union and just emerging on its own. The Soviets packed up and left, leaving the locals with a lot to figure out for themselves. And so we chartered a plane out of London with pilots who were experienced in the region. We had Russian visas because we were going to visit the city of Perm in Russia, which was a large oil-refining center. So, we flew into Tashkent, and because private planes were rare, the Uzbekistan authorities let us into the country without any hassle.

The customers were happy to see us and gave us a nice dinner. And when our visit was over, we headed back to the airport where a representative directed us to go to a room and wait for the customs agent to process us out. But hours went by, and no agent appeared.

Finally, someone came in and told us he was on his way. When he finally arrived, we knew what took so long. He was drunk and needed some time to sober up. We handed him our passports, and his reddish face grew even redder. He said, "These are Russian visas and Uzbekistan is not part of Russia anymore. Where are your Uzbekistan visas?" We didn't have any. Because we'd flown into Uzbekistan on a private airline, we hadn't officially checked in to the country.

This bureaucrat was now faced with the biggest decision of his life, and he had a total look of consternation. In my mind, I thought we were caught in some unbelievable twist of red (no pun intended) tape.

But after fussing for a while as his hangover started to subside, his eyes lit up and he said, "I know what to do. You were never here! Have a good day!" Not quite believing what we'd just heard, we didn't stick around to ask questions and hightailed it out to the plane. The pilots quickly fired up the engines, and we took off, happy to see the Uzbek airport fade away under our wings.

While that experience gave me no shortage of gray hairs, I'm still glad we did it, and it reiterated a critical point in my mind: *customers, regardless of where they are located, are always first.*

MAKE STRONG ACQUISITIONS

One of the best ways to strengthen your position in the marketplace is to make strong acquisitions. Sometimes this works out well, and other times it does not.

As mentioned earlier, a rigorous competitive analysis is going to reveal gaps between you and the competition. These gaps might include areas such as technology, global reach, expertise, or market position. In my career, I was involved in many acquisitions. Some of them didn't meet the expectations we had when we proposed them. Others turned out even better than we thought.

Traditional acquisition justification revolves around a projected business plan. In most cases, there are three plans that include a base case, a downside case, and an upside case. Each case outlines both revenue and cost synergies as well as potential savings in areas like time to market. We always went for acquisitions that were successful in their own right and avoided "fixer-uppers."

An acquisition that was successful beyond our expectations was the 1999 acquisition of the Westinghouse Process Control (WPC) Division. This division was owned by CBS (yes, the broadcasting network). At the time, Westinghouse Process Control was very successful with control systems focused on the power, water, and wastewater industries. Their in-house experts knew these industries inside and out.

At Fisher-Rosemount Systems, we had very little penetration into those industries. CBS decided to sell Westinghouse Process Control and solicited competitive bids. We wound up being the successful bidder. The employees at WPC were happy to be part of a business that focused on automation, and so were their customers.

We gave this business the charter to focus on the power, water, and wastewater industries, and we looked for ways to leverage each other's technology. The business took off, driven by the secular trends in the power industry of replacing coal-fired plants with natural gas and adding alternative energy sources, such as wind and solar. Today, this business offers grid management, giving their customers the ability to integrate all their sources of power generation. And as a result, it is well positioned for the future.

An acquisition that didn't work out so well was one we made in the mid-1990s. If you'll recall, we wanted to use PCs as operator stations for DeltaV. And these operator stations have visual displays of the plant, such as a tank with the changing level shown by color on the screen. These colors would show the flows and the ability to pop up alarm conditions to alert the operator. And there was a company called Intellution that had developed what was called a human-machine interface. In other words, the graphic software would enable an engineer to easily configure displays rather than writing them in code.

Intellution was founded by Steve Rubin, one of the smartest entrepreneurs I've ever met. Rubin also had the vision to use PCs as operator stations and built his business on the concept of software that ran on a PC but could be used on systems called programmable logic controllers (PLCs).

PLCs were primarily used in the discrete manufacturing business with anything from automobiles, to filling cans with liquid, to controlling sequences. Intellution was a leader, and we saw an opportunity to drop their software into DeltaV's PC-based operator station. This would save us time to market and considerable development expense.

What we didn't fully understand was that Intellution needed to be seen as totally agnostic about the system it was paired with. And as a result, we found ourselves in competition with them. Their sales organization wanted to sell the Intellution software and did not push any system. But when DeltaV was bought by a customer, the DeltaV operator station was included with the system, and the Intellution salesforce did not make a sale.

This experience taught me an important lesson: when considering an acquisition, spend some time on the possible unintended consequences. In the end, we ended up selling Intellution to GE.

THE MAIN COMPETITIVE GIANT OF TODAY

The giant of competition looms larger today than it's ever been. Even small businesses compete with global competitors.

There are several examples we could use, but one that stands out is the rise of artificial intelligence (AI). With the surge of new AI tools, we've reached a new technological tipping point. And while the worlds of the mid-'90s and the mid-2020s feel worlds apart, the same principles for competition I used three decades ago can still apply today.

Many are quick to talk about the downsides of AI, and there are some potential negatives. Not the least of these are the increased possibilities for terrorism. But with these increased risks are fresh opportunities. Some of these include the following:

- **MEDICAL**—With enough advancements, AI will help patients receive complex diagnoses the average family doctor might overlook. There is also hope for options such as precision surgery, reduced wait times, and limiting the risk of human error.

- **DENTAL**—Several months ago, I needed a crown for my tooth. In the old days, they used to make a mold, and I'd have to go back to the office to get my crown placed several weeks later. But now, the dentist placed a camera probe in my mouth and used a 3D printer to make my crown. Forty-five minutes later, I was all set. These advancements will only get better.

- **TRUCKING**—The average semi-truck gets around 6.5 miles per gallon. Unfortunately, because of poor scheduling, many of these trucks end up running "empty miles" upward of 20 percent of the time. AI could help reduce this number substantially.

- **AUTOMATION**—There is a huge opportunity for AI to improve production, lower energy consumption, reduce emissions, and improve safety.

- **AUTOMOBILES**—Drivers already have an array of driver assistance features, and AI will help them improve even more.

The list could go on forever because the possibilities for AI are endless.

What we're talking about here is the convergence of information technology (IT) and operational technology (OT). When information merges with operation, real innovation occurs.

Back in the '90s, most companies feared opening their control systems to the internet. What if a competitor or someone with bad intentions gained access and wreaked havoc on the entire system? As a result, control rooms were isolated. And anything uploaded was done on a private network. Data never left the confines of the facility.

The manufacturing process that controlled what a company made tended to be an island. As such we called it the "island of automation." Now, everything is completely different. Companies upload to the cloud, and data analytics are applied to this data to mine for any kernels of knowledge that might help the system run more efficiently.

Now, with "machine learning," systems can learn from their mistakes and improve. One example of this was the series of games IBM's Deep Blue computer played against world-renowned chess champion Gary Kasparov in 1996–1997. Because the computer could learn from its mistakes, it was able to get better and better. A more recent example occurred back in 2015–2016 when a company called DeepMind Technologies created a computer program called AlphaGo to play the popular boardgame Go.[25]

Because Go has so many more possibilities than chess, skeptics wondered if AlphaGo could even compete. But as it turned out, it was more than able. And in March 2016, in a series of five games, it bested top-ranked Go player, Lee Sedol, 4–1, leaving Sedol to say, "I apologise for being unable to satisfy a lot of people's expectations. I kind of felt powerless."[26]

---

25 Google DeepMind, "Introduction to Artificial Intelligence," YouTube video, March 15, 2022, https://www.youtube.com/watch?v=WXuK6gekU1Y&ab_channel=GoogleDeepMind.

26 Jim Edwards, "See the Exact Moment the Retiring World Champion of Go Realised Deepmind's Machine Was 'an Entity That Cannot Be Defeated,'" Business Insider, November 28, 2019, https://www.businessinsider.com/video-lee-se-dol-reaction-to-move-37-and-w102-vs-alphago-2016-3.

For many, they resonate with this sense of powerlessness that Sedol felt. They look at how much technology has changed in the past five years and dread the thought of a new evolution.

Believe me, I get it. But if you plan to be in business longer than ten years, it's critical that you embrace a healthy form of paranoia. New AI sites are popping up every minute, and the real question leaders should be asking is, "How can I gain a competitive advantage by leveraging the power of AI to my advantage?"

Rest assured your competitors are already asking this question. And rather than see this moment as a time to shrivel up in fear, see it as an opportunity to give your competitors a taste of the innovator's dilemma.

CHAPTER 6:

THE GIANT OF SUCCESS

Of all the giants to overcome in this book, this is the one I enjoy writing about the most. Everyone wants to be successful, and most would be content to have this be the *only* chapter in their life's story. But turning the giant of success was only possible for me *after* I turned the giants of corporate bureaucracy, doubt, innovation, skepticism, and competition. Without these turns, success wouldn't have even been an option.

Sometimes we throw around the term "overnight success" to refer to someone who makes a large sale or lands a massive contract. But what most of us fail to see are the thousands of hours that precede this breakthrough. We don't comprehend the sleepless nights, endless meetings with management, or hundreds of unsuccessful pitches to potential customers.

Only after studying the details of someone's life do we begin to realize that one of the distinguishing marks that separates successful people from unsuccessful people is a person's ability to thrive in the mundane. Successful people are able to stick with something when others give up, and they keep pressing forward even when they don't always see measurable progress.

In the words of Jim Rohn, "Successful people do what unsuccessful people are not willing to do."

THE ANATOMY OF A SUCCESSFUL LEADER

To be successful, you must put in the time. There are no shortcuts. You must lead by example and show your team you're working for them. You're not just asking them to make your organization better, but you're actively helping them hit their career goals.

Unfortunately, now more than ever, it's easy for CEOs to hunker down in their offices and hide behind written reports and Zoom calls. While each of these actions have their place, they are no substitute for showing up in person. No report or video call can convey the scope of human emotion. And there is simply no substitute for being present.

Real leaders are individuals who not only lead the charge but are also willing to get their hands dirty. They're not like some of the political figures we see today who base their decisions on getting reelected. Instead, the best leaders make decisions by doing what is in the best interest of those they lead.

One of my favorite leaders was President John F. Kennedy. I was only in the eighth grade during his run for office in 1960, but even at that young age, I still remember being awestruck by the way he communicated. And as I grew older, my admiration for him only increased. His ability to inspire and take charge of any situation left an indelible imprint on my life.

Several years ago, I visited the Kennedy Library in Boston. And on one of the desks, there was a NASA report that outlined the plan for placing the first man on the moon. What struck me was not the main text on the paper but Kennedy's notes he'd scribbled in the margins. There must have been forty or fifty handwritten thoughts like "too much money" or "needs to be quicker."

As I looked at that paper, I thought to myself, *it's no wonder I liked this guy. Here was a leader who knew how to get things done.*

Kennedy often spoke without notes but still provided detailed and knowledgeable answers. His wit was powerful. I think of one occasion where a reporter, who was less than happy with his work, stood and asked a question. "Mr. President," she said, "the Democratic platform, on which you ran for election, promises to work for equal rights for women, including equal pay to wipe out job opportunity discriminations. Now you have made efforts on behalf of others. But what have you done for the women according to the promises of the platform?"

Taking this chiding question in stride, Kennedy smiled and said, "Well, I'm sure we haven't done enough."[27] The crowd of reporters laughed, and Kennedy then went on to share some of the steps he *was* taking. With this little twist of humor, Kennedy turned a potentially awkward interaction into a pleasant exchange.

Ronald Reagan was another brilliant communicator I admired, in part because he'd experienced success as both a politician and an actor. He knew when it was time to get serious and tell Mikhail Gorbachev to tear down the Berlin Wall. But he also recognized when to break the ice with a well-timed joke.

According to Reagan, "The greatest leader is not necessarily the one who does the greatest things. He is the one that gets the people to do the greatest things." And when I think of leaders who get

---

27   e2films, "The Wit of JFK," YouTube video, November 25, 2015, https://www.youtube.com/watch?v=BRcTCUTXr5M&ab_channel=e2films.

their people to do great things, I think of modern-day examples like Volodymyr Zelenskyy, who are willing to stand by their convictions, even when doing so might cost them their lives.

SUCCESS COMES IN MANY FORMS

I like to think of success in terms of impact rather than popularity. With our modern media age, we have a host of celebrities I'd call famous but wouldn't necessarily label a success. Real success is impacting people around you for the better. Sometimes this happens on a global stage, but most times this occurs in ways only a few observe.

When I think of lesser-known heroes in my life, my mind immediately goes to my high school math and drama teacher, Mr. Joe Schulte. He was a renaissance man who could write complex calculus equations in the afternoon and teach drama after school. In fact, it was because of him that I went into drama.

And when I was part of Mr. Schulte's acting group, he did everything he could to help his students understand the real art of drama. Every year, he would purchase tickets to a Broadway show in St. Louis and take a group of students to attend. And the next day after the show, we'd analyze everything that was said. *Why did that character say this? Why did this person wear that? What was the major theme of the drama, and what were some of the hidden meanings?*

I loved it. However, I'll have to admit I was much better at math than I was at drama, a point Mr. Schulte went out of his way to remind me of at my fiftieth high school reunion. When I think back on this interaction, it was such a moving experience to see him again and tell him what he meant to me. Sadly, he passed away a few years later, but the skills I gleaned about stage presence helped me when I went into business. I learned how to project confidence, speak with clarity, and share my message with conviction.

To me, Mr. Schulte's life was a success because he lived it in service to others and made a positive impact in the lives of the young people in his classes. While few people know his name, I admire this man more than any professional businesspeople I know. He was a success. And while success can be defined in many ways, I like what author Mark Batterson wrote when he said, "Success is when those who know you the best respect you the most."[28]

EXPERIENCING PUBLIC SUCCESS

While I experienced quiet successes throughout the 1970s and 1980s, it wasn't until 1993 that my career really took off.

From 1993 to 1997, I served as president of Fisher-Rosemount Systems. During this time, I merged and restructured two companies into one $400 million entity, grew sales at 10 percent compound annual growth rate (CAGR) from 1994 to 1997, improved profits from loss to 12 percent operating profit, and grew non-US sales from 50 to 65 percent of total sales. I also introduced the industry trans-forming DeltaV PC-based control system and received a Technical Innovation Award from Microsoft and best vision rating from the Gartner Group.

Up until 1995, Joe Adorjan was president over all of Fisher-Rosemount. This included systems, valves, and measurement. But then Joe moved on, and his job became available. I wondered if I might be considered for this role but quickly dismissed this thought. I hadn't accomplished enough yet, and the jury was still out on whether

---

28   Mark Batterson (@MarkBatterson), "Excited to Announce the Release of My New Book! #Book-Launch," Twitter, June 15, 2023, https://twitter.com/markbatterson/status/1140287974267129856.

DeltaV would be a success. As it turned out, Bill Davis became the new president. This didn't surprise me as I figured he would replace Chuck Knight as CEO of Emerson one day.

But within a year and a half, Bill moved on, and now I thought I was ready. Instead, a man named David Farr got the job. Initially, I was disappointed. David wasn't an automation guy, but we got along well. David was an energetic leader and gave me a lot of support. And as it turned out, I didn't have to wait long for my chance. Chuck Knight retired as CEO of Emerson in late 2000, and David took his place.

This opened the door for me to become president of Fisher-Rosemount and executive vice president of Emerson. When I took over as leader of Fisher-Rosemount, the whole of Emerson was around $25 billion. Fisher-Rosemount was a $2.7 billion business and one of five major businesses under Emerson. And within Fisher-Rosemount, the systems business was small (around $400 million) in comparison to the measurement and valve businesses. This meant I was stepping into a role that was more diversified and seven times larger. It included multiple companies on both the measurement and valve side of the business.

Getting the job as the leader of all of Fisher-Rosemount and becoming an executive VP of Emerson was my dream job. I was in this role until 2008, became chairman of the group in 2008, and retired in 2010. I grew sales from under $2.7 billion to $6.7 billion, thirty thousand employees, earnings before interest and taxes (EBIT) from 11.2 to 19.7 percent, and non-US sales from 56 to 67 percent while still making penetration gains in the United States. This included 45 percent of sales in emerging markets. And I helped acquire over $1 billion in global acquisitions and achieved the #1 position in a significant number of automation categories, as measured by the Control Magazine Readers' Choice awards.

But even though my entire career had been in automation, I had a lot to learn.

BRINGING COHESION

One of the most important steps I took after becoming president over all of Fisher-Rosemount was to bring cohesion to the various companies that fell underneath our umbrella. Because Fisher-Rosemount was under the Emerson brand, sometimes it got confusing when Fisher-Rosemount acquired another company like Micro Motion. Now we were several layers deep. It was confusing for customers, and these acquired companies sometimes felt like third-wheel operators. They were neither Fisher nor Rosemount. As a result, our business felt fragmented.

Automation customers needed measurement, control, and valves to accomplish their mission. But while all of Fisher-Rosemount had a common customer base, we were flooding each customer with multiple business card formats, different warranty processes, and different credit terms. There was no continuity.

My goal was to harness the power of the entire group and eliminate some of the branding confusion. To accomplish this, we changed the name of Fisher-Rosemount to Emerson Process Management. Practically speaking, this meant that when someone from a company like Micro Motion handed their business card to a client, the card might say "Micro Motion" at the top, but the bottom would include "Emerson Process Management." We also went through the delicate process of standardizing a common format for business cards, stationery, advertising, and signage.

A good example of an umbrella brand is Proctor and Gamble. Under their umbrella, they have numerous baby care, fabric care, and body care companies. While each has its own name such as Gillette, Bounty, or Tide, the goal was to form some sort of cohesion so that each company worked for the betterment of the entire whole of Proctor and Gamble. This is what I worked to accomplish when I took over as president of Fisher-Rosemount.

We were one major automation organization and not a federation of companies. I wanted each company under the Emerson Process Management umbrella to be working together with the other companies and not feel like they were competitors. And so we streamlined our salesforce and moved toward a more unified sales management for Emerson Process Management.

As the years have progressed, this cohesion has only increased. Today, Emerson Process Management is now simply called Emerson because Emerson has focused its business on both process and manufacturing solutions.

MY GREATEST SUCCESS

Along with cohesion in branding, I wanted cohesion in the products and services we offered to our customers. And one of the ways I brought everything we offered together was through the innovation of an ecosystem called PlantWeb.

I received the first glimmer of this futuristic concept while I was still working at Rosemount. And during one of our sales meetings, I presented what I thought was the future of the instrumentation and control business. What stimulated my far-out thoughts was a 1968 movie called *2001: A Space Odyssey* starring Gary Lockwood,

Keir Dullea, and a third character that was not human—a computer named Hal 9000.

If you've watched this film, you know that Hal eventually turns on his two human counterparts, causing a series of drastic measures. The movie predicted many things that are commonplace today, even AI. The value of movies like *2001: A Space Odyssey* is that they made us think into the future. And while some of the decades-old projections people made about technological advancements have proven false, many have become reality.

And as I stood in that small office and thought about what the next few decades might hold, I envisioned a world where instruments in the field would do more than measurement. While I certainly didn't have any formed ideas about what this might look like, I was confident in the concept of more and more diagnostics and being able to send digital information over the same wires used in automation.

And over the next thirteen years at Rosemount, I saw gradual movement in this direction. But it wasn't until I became president of Fisher-Rosemount Systems that everything started to come together. Then, when I took over as president of Emerson Process Management, I continued to push the theme of innovation. And the biggest innovation was to pull all the products together into an ecosystem called PlantWeb. This is where we drove a common "look and feel" in the setup software used to configure everything. We also pushed the concept of asset management.

This was made possible because new breakthroughs continued to happen with devices in the field. As they continued to get smarter, they were able to communicate accurate projections to the control system. Along with this, the technology in personal computers grew by leaps and bounds. And third, digital memory became dirt cheap. Today, we take for granted having an iPhone with 250 GB. But in the old days, everything had to be coded as efficiently as possible to save space. Now, everything was different, and our field devices became much more sophisticated.

And by building sophisticated diagnostics into the devices in the field, we could send status messages over the standard digital or hybrid field communication protocol back to the control room. We developed a suite of asset management software that could run either in the DeltaV operator station or in a stand-alone PC.

Let me give you an example. When the controller sent a signal to the control valve to go to 50 percent open, a smart device on the control valve would check the valve position and tell you if the control valve actually went to 50 percent. Stuck valves were very bad for the process and for safety. This one diagnostic solved a major customer pain point.

DeltaV continued to innovate and added a safety system. This system operated independent of the control system and was loaded with redundant algorithms to achieve maximum reliability. Its purpose was to use measurement products (separate from those used in control) to detect an unsafe condition and shut the plant down in an orderly manner.

The basic concept behind PlantWeb was to tie our smart instruments, valves, and systems together in a way that utilized full digital communication to the control room and back again. This meant a whole new layer of things like diagnostics could be generated.

For example, we could tell through having measurements of vibration on a compressor that a bearing was starting to go out. This was asset management combined with process control. The concept was both simple and powerful. Each product from Emerson Process Management was best in class and could stand toe to toe with any of their individual competitors. But when they were all put together into one automation package, an entire additional level of customer benefit came about.

I'd pushed this concept before I became president of Emerson Process Management. But when I got the whole job, I pushed it harder and had the ability to force certain issues. I had so many more resources at my disposal. By the early 2000s, we had PlantWeb going and had PlantWeb University where customers could go and take courses. We offered graduation certificates, and the whole experience was somewhat unique. PlantWeb sent a signal to customers that we were more than just process control.

When customers asked me to explain PlantWeb in simple terms, I asked them to envision having a watch that told them their cholesterol levels. Apparently, I was ahead of my time. But my point was that just as an Apple Watch today can let people know their heart rate, PlantWeb could diagnose the heartbeat of a manufacturing facility and predict problems that might occur.

The value of this product hit home to me one afternoon when I met with a man who oversaw all of Shell's refineries in North America. I asked him to share his greatest challenge.

Without hesitation, his head jerked up and he said, "Unplanned shutdowns."

And for the next few minutes, I shared with him the basics of PlantWeb. I told him how this system could not only diagnose problems but also that our predictive algorithms could send alert messages before a problem became serious.

He grew excited, and over the course of time, we revamped several refineries with PlantWeb. We were leveraging all our strengths at the same time. And this soon propelled us into becoming the chosen supplier on many projects.

The impact of this innovation was equally good for the company. We grew like crazy and made more acquisitions. And the impact this had on the competition was massive. It was another one of those things that wasn't easy to copy.

For example, some of the big system companies didn't have *any* control valves. And this was a big deal because control valves are rich in diagnostics. In fact, the entire process flows through the valves. And as the valves get exposed to temperatures and pressures, they wear out. Before PlantWeb, the maintenance practice on valves was just time based. When it was "time," the process was shut down, and valves were taken out of service, inspected, and repaired if necessary.

This was called *preventive maintenance*. But with PlantWeb, our smart valve monitored performance over time. It watched how much force was needed to move a valve, and when the force got past a preset limit, a message was sent that the valve was wearing out and would need repair. This enabled our customers to practice *predictive maintenance*, which was less expensive and less disruptive.

PlantWeb's launch was a chance for us to take an innovative approach to marketing. Around this time, we hired a consultant named Kathy Button Bell. Kathy was a creative dynamo who helped us produce ads that made us stand apart from what typically appeared in trade magazines. These were called metaphor ads. Kathy also placed ads in airport kiosks in key customer cities, and she skinned the buses that provided transportation to and from hotels at trade shows.

Today, Emerson advertises PlantWeb on CNBC. And while there are suppliers who now offer asset management, PlantWeb remains ahead of the game.

THE COST OF SUCCESS

This is not to say the successes I had didn't come without costs to me and my family.

The other day I was watching an interview with Elon Musk, and he talked about the whole "work from home" culture in America. This might work for some people, but it certainly was not an option for me. Throughout the 1980s, 1990s, and early 2000s, I was on the road all the time.

During the late '90s, I reviewed my calendar at the end of the year and noticed I'd only spent 25 percent of my time at the office. The rest was on the road. I was constantly meeting with other industry leaders, clients, or our team members who lived in different parts of the world.

It was Woody Allen who said, "Ninety percent of success in life is just showing up." I had a philosophy that I would visit my customers, regardless of where they lived. And by regardless of where they lived, I'm not talking about a trip to a remote part of the United States.

I say this because we live in an age where a mere 32 percent of workers are engaged at their place of employment.[29] This tells me few are willing to pay the price for success. Certainly, during the depth of COVID, people had to stay home. There were so many uncertainties. But those days have thankfully passed. And to be a leader of a company, you must be willing to see your customers whether they be in New York or Indonesia.

Had I sat back and just read the reports from my office, I wouldn't have been half as effective as I was. I had to visit facilities, go through sales offices, answer questions, and be a leader on the ground.

Sometimes these experiences were far from pleasant. I got food poisoning multiple times, and there were moments I didn't know

---

29   Jim Harter, "Employee Engagement Slump Continues," Gallup, September 3, 2023, https://www.
gallup.com/workplace/391922/employee-engagement-slump-continues.aspx.

if I'd ever be the same again. There were times my doctors couldn't figure out what was wrong with me because I'd encountered a disease they weren't accustomed to treating. I still recall the time one of my doctors said, "John, I don't know what's wrong with you, but I know you're sick."

No duh, I thought to myself.

"The best suggestion I have for you at this time is to hook you up to IV and flush whatever this is out of your system," he continued. I just nodded, too miserable to put up much of a fight. And in a few weeks, I was back to normal.

But stories like this just go to show some of the personal price tags that came with my success. During many of my trips overseas, I traveled a lot with David Farr, and he had the same point of view about on-the-ground leadership. He wanted to be visible and be there to dedicate new facilities. We both saw this as the obligation of a leader.

The mental strain of my various positions was also high. Because I never let my guard down in front of my team, this meant I carried a lot of the pressures home. And as a result, family sometimes paid the price.

I recall a conversation my youngest son had with Charlotte, telling her, "Mom, I don't want to do what Dad does because he's always at work." Ironically, this son, who is now a man, is one of the hardest workers I know. But at the time, his words stung because there was a part of me that couldn't argue with him.

The reason I share these personal points is to underscore that success is messy and always comes with a cost.

THE QUALITIES OF SUCCESSFUL LEADERS

This cost of successful leadership includes a daily commitment to good disciplines. Unfortunately, sometimes leaders rise to the top of their organizations and start thinking of themselves as omnipotent. It's the whole "absolute power corrupts absolutely" dilemma. And the moment this happens is the moment the organization starts to unravel.

But one of the real joys of my forty-two years in the automation industry was working on leadership development. It was a treat to see people grow. And from firsthand observation, I realized leaders show up in many different forms.

Sometimes, when we think of leadership, we tend to picture the president of a country or a company CEO. But leadership is something that we all need, whether we are a parent, a member of a volunteer group, or a company executive. If you think about it, it is rarely possible that good things come about because of just one person acting alone. In the sea of competencies needed for leadership, I have found that there are six that stand out. These include vision, passion, integrity, action, influence, and belief in others.

Let's look at each.

VISION

First, vision is not really a crystal ball that projects way into the future. Vision is looking around the corner. It is the ability to sort through complexity and see a pattern and then project that pattern into a future picture. Vision is not only about the picture, but it is also about painting that picture in vivid terms and "selling" that picture to those whose efforts are needed.

Vision requires us to develop a certain amount of comfort with ambiguity. It acknowledges there will be setbacks, but that we must be like the inflatable clown that keeps popping back up.

One of the greatest examples of leadership through vision was Lee Kuan Yew, the first prime minister of Singapore. Facing a devastated postwar Singapore, Lee's vision was a city-state that would not only survive but also prosper. He articulated his vision very simply by stating, "We are going from third world to first."

From that vision sprang several beliefs, which he continuously pushed. Water had to be safe to drink, politicians could not be corrupt, and he would favor businesses that needed knowledge.

Because of Lee's efforts, Singapore is the thriving nation it is today. He saw a vision and pursued it with relentless abandon.

PASSION

This brings us to passion.

Passion is not about pounding a table or raising your voice. Passion is showing commitment in every meeting, in every one-on-one conversation, and in every decision. People see through phony passion easily. Passion is about focus. It is demonstrated by our actions and the questions we ask. Passion means having courage to stand up for what we believe.

My favorite example of passion is Winston Churchill. He had strong beliefs about the impending danger of war in Europe. England turned to him in the crisis. If you want to hear passion, go online and listen to his speech after the retreat at Dunkirk. Listen to the passion in his voice. He wasn't pounding a table, but his speech inspired the country to rally and never give up.

I think back to the previous example I shared of Volodymyr Zelenskyy. Here was an actor and comedian who became the unexpected president of Ukraine. But like Churchill, he has united his country despite the hardships, and his speeches and addresses to Western countries have motivated them to support him in Ukraine's fight against the Russian invasion.

INTEGRITY

Integrity is essential in a leader. When most of us think of integrity, we think of the obvious things like telling the truth and owning up to our mistakes. That's just the beginning. Leadership integrity means that you don't set one set of standards for your people and a different set for yourself.

Leaders must earn trust with unwavering daily demonstrations of personal integrity. The fastest way I know to lose trust is to walk around like you're God's gift to the world. A good friend of mine always used to say, "If you are the Pope, you have to be the most Catholic of all."

One of the greatest examples of leadership through integrity in my life is Vern Heath, the founder of Rosemount. He was revered and loved by employees at all levels. This is not to say he was a teddy bear. He certainly wasn't and could be demanding and hard-nosed. But employees loved him because his own behavior was unwaveringly consistent with the values he expected from his employees. Not once did I see him display his ego or exempt himself from the rules. When tough times came, the employees accepted his direction and understood the need for austerity. He had the total trust of his employees.

How can you follow someone whom you can't trust?

ACTION

Action in the leadership context is not just motion. Action means driving for results, making decisions, and challenging the organization to get its blood pumping.

In my experience, there is nothing worse for an organization than uncertainty. When people don't have a clear picture, they understandably fill in the blanks with the worst possible scenario. The fear of making a mistake holds many leaders back from making a clear decision. What they don't realize is that inaction is the biggest mistake they could make.

In my career, one of the greatest examples of action was Bud Keyes. I mentioned Bud's role in the development of DeltaV. Bud was a dynamo. He got by with very few hours of sleep each night but would outwork everyone else. He was a voracious reader. Bud developed a brain tumor, but that didn't stop him. Amazingly, I got an email from him right after he got a radiation treatment, giving me a detailed technical description of how the process worked and hoping that the software in the device was robust and bug-free.

INFLUENCE

Influence is the most important and least understood competency. It is not politics or connections. It is the ability to motivate and energize others. In the words of John Maxwell, "Leadership is influence, nothing more, nothing less."

Influence means getting people's hearts before you get their minds. People must believe in you, or they won't follow you.

One of the most powerful tools for influence is to ask people to project into the future and visualize how they will feel when they look back on themselves. Will they look back with regret on missed opportunities and half-hearted efforts or in joy at knowing that this was their finest hour?

There is no more powerful motivator than anticipating how sweet success will feel.

BELIEF IN OTHERS

Belief in others may seem obvious, but many leaders fail because they don't stand behind the team.

Real leaders do not take all the credit for their success, and they don't run for cover when there's trouble. They believe in their people and seize every opportunity to develop new leaders. And they look for opportunities to praise what those on their team are doing. They do not have a blind belief in people and understand the difference between carrying the wounded and saying goodbye to stragglers.

Influence and belief in others go together. Think about the people who believed in you and were profoundly influential in your life. It could be a relative, a teacher, or a friend. This is leadership at its essence.

STUMBLE TOWARD SUCCESS

Ultimately, successful leadership only happens as you learn to turn your giants. And my hope is this book has inspired you to think through the complex obstacles in your life and rethink the way you lead. Instead of looking at each challenge as a problem to eliminate, you see it as a giant to turn.

Winston Churchill said, "Success is stumbling from failure to failure with no loss of enthusiasm." And this quote is the summary of my life. I've had too many failures to count, but the key was I kept moving forward even when I couldn't see the path.

It's natural to either run from giants or try to mow them over. But it takes a lot more patience and fortitude to methodically turn them to your advantage. However, when you look at the lives of successful people you admire, rest assured that this is something they learned to do.

CONCLUSION

My goal in writing this book wasn't to create an extended bio of my personal accomplishments but to hopefully offer perspective. Sometimes when you're in the thick of the daily grind, it's hard to step back and see a path forward. All you see are the problems and challenges you face. But I want you to know there is hope.

Turning your giants doesn't happen by accident. You can't snap your fingers and make them disappear, and it isn't helpful to run from them. Instead, turning your giants requires intentional pursuit and persistence. And as I often say, persistence is passion applied. When you are truly passionate about something, you will find a way to make it happen.

But like the inflatable clown that gets knocked down and keeps popping back up, you need to develop resilience. When one giant knocks you down, get back on your feet. I cannot stress how difficult it is to think and live this way. When many people get hit, they stay down for a long time. Rather than developing their mental fortitude, they cave to the pressure and quit. And the end result is a life of regret.

Now, just to offer a point of clarification, not everyone is a giant-turner in the same way. I think of my parents as key examples. My dad was willing to engage in combat and fight in some of the toughest battles in World War II, but there were points he pulled back from a giant when I might have tried to engage.

But if there was anything my dad taught me, it was to not live a life of regret. There was one specific example he often shared.

After finishing high school, Dad went to work in a brickyard alongside my grandfather. This was during the Great Depression, and it was obvious his family needed the money. But my dad was also a terrific baseball player and played on an American Legion baseball team in St. Louis.

One day his team wound up playing a game in the old Sports-man's Park, home of the St. Louis Cardinals. Dad played shortstop and made some wonderful plays in the field and tripled off the wall.

There was a New York Yankee scout in the crowd, and he was impressed with Dad's combination of blinding speed, agility, and hitting power. After the game, he came to Dad and asked if he could come to his house to talk to his parents about a possible minor league contract. Dad readily agreed, and the scout visited his home and explained to my grandparents how he would like to sign their son to a contract.

But my grandparents were immigrants and could not understand why anyone would pay someone money to play a game. They said no, and that Dad must not give up his job in the brickyard. Dad obeyed and was not signed.

A few years later, an identical scene played out in another Berra household in St. Louis. And another young man named Lawrence Peter Berra, no direct relation to my family, signed a contract with the Yankees. Of course, we know him today not as Lawrence but as Yogi.

As I've thought back on that story, I have a mixture of emotions. On one hand, I can understand why Dad didn't make this decision. He needed to help provide for his family. But on the other, I know it's a regret he always had, and I'm sure he never stopped wondering what might have been.

When I think about my own children, one of my encouragements to them is to never live a life of regret. In *Top Five Regrets of the Dying*, Australian nurse Bronnie Ware listed the five regrets her dying patients expressed to her shortly before they passed. These regrets included the following:

- I wish I'd had the courage to live a life true to myself, not the life others expected of me.
- I wish I hadn't worked so hard.
- I wish I'd had the courage to express my feelings.
- I wish I had stayed in touch with my friends.
- I wish that I had let myself be happier.[30]

Not every giant is yours to turn. There is always a trade-off with some sacrifices attached. So choose your battles wisely. Know where you want to go and then do whatever it takes to get there. Understand that titles are not your rightful inheritance just because you graduated from the right university. And recognize you must prove yourself every day.

KEEP DISRUPTING

One of the ways you do this is by developing the mindset of a constant disrupter. Rather than sitting back and resting on your accomplishments, you keep moving toward the next thing. You avoid compla-

---

30   Bronnie Ware, *Top Five Regrets of the Dying* (Melbourne: Bolinda Publishing Pty Ltd, 2017).

cency like the plague and are paranoid about what your competition might be doing.

Remember, the moment you stop thinking like an innovator is the moment your company will fall behind. If you're sitting atop the competition, be careful about what sort of disruption can happen to you. Think about how you can disrupt your industry again. Likewise, if you're on the bottom looking up, keep in mind that your ability to move up the ladder is contingent on your willingness to disrupt and be innovative.

Innovation is all-encompassing. It's not just about the product or service. Innovation touches everything from your websites, social media interactions, and the way you manage your brand. And leaders must promote an innovation culture even if they aren't themselves innovators.

KEEP PAYING THE COSTS

As you continue to disrupt, don't forget to pay the costs. In the words of blues legend B. B. King, "you've got to pay the cost to be the boss." Unfortunately, this is a lesson many leaders have forgotten.

During my leadership career, I went on countless sales visits to support our salesforce. Most of these experiences were positive. But it was clear to me that many had disdain for anything I said. There was this notion that all my talk about automation was just a used-car-salesperson pitch. The belief was, one supplier was just as good as the other, so why waste time looking at different options?

I think of one interaction I had with the head of a large oil refinery. I'd been asked to visit his office to advise him on the specifics of PlantWeb and show him how this could help his company's operating performance. But from the second I walked into his office,

I knew something was off. Every few seconds he'd glance down at his watch, and it was obvious he couldn't wait to get rid of me.

When I finally stopped and asked him if there was a problem, he said, "Frankly John, as far as I'm concerned, all automation projects have zero return on investment." Translation: *Your presence here is irrelevant. And so is the presence of my own automation engineers.*

I couldn't believe it. I thought, *You dunderhead. Nothing could be further from the truth! Don't you want to make your refinery safer? Don't you want a higher-quality product? Don't you want to grab a larger share of the market? Automation is usually no more than 7 to 10 percent of spending when you're building a new plant. And if it works, that plant will sing!*

Of course, I didn't tell him any of that. He was beyond hope, and so I ended my presentation and walked out.

This was a person who thought talking to anyone from a vendor organization was beneath him. In his mind, he had engineer minions to deal with this stuff. But the moral of this story is that every big company has automation engineers in its workforce. However, they often labor in obscurity when, in fact, they are the unheralded keys to greater growth and profitability.

The ironic point of this story was this refinery was part of a larger organization that owned multiple refineries. And I knew from the numbers that one of the other refineries that used PlantWeb was doing very well and that their numbers were much better than the refinery I was visiting. And I had no doubt that part of the reason for this was because of the genuine operational improvements we offered. Unfortunately, the character of the person in charge of this operation meant his refinery lagged behind.

I've had this same type of conversation numerous times. During my years at Rosemount, I recall a time when I was speaking about transmitters with an executive. As we spoke, he held up his hand and said, "You know what I think? I believe transmitters are like toilet paper. One brand is as good as the other."

At the risk of losing this account, I said, "Well, you don't notice the difference in toilet paper either until you get hemorrhoids."

My point was that while the cost differential in transmitters was comparable, the quality was not. And purchasing the wrong transmitter could result in disaster for the entire plant operation.

From my vantage point, if the top management of an organization has the attitude that they can't get into some of the details of what is running their business, God help that business. Each decision will be made based on a financial decision with little thought given to the overall quality of the product.

KEEP YOUR COMMITMENTS

As you embark on this turning the giants' journey, keep in mind your need to have personal integrity. No matter how much you might think about it, you are not God's gift to the world.

Don't demand from your people something you don't demand from yourself. Do what you say you're going to do. Follow through on your commitments. Remember that integrity generates trust and people won't follow you if they don't trust you. According to Brian Tracy, "The glue that holds all relationships together—including the relationship between the leader and the led—is trust, and trust is based on integrity."

We live in a world where there seems to be a new case each day of a popular leader who broke the trust of those he or she led. And as a result, this has caused many to mistrust those in positions of power. I urge you not to add your name to this list. Remember that trust takes a lifetime to earn and a day to lose.

So keep your commitments and lead your people well.

KEEP TURNING YOUR GIANTS

If you haven't already, I'd encourage you to take out a notecard and list some of the greatest giants you face today. Write down what makes them so challenging. And start thinking of creative ways you can turn them to your advantage.

Do you struggle with the giant of corporate bureaucracy? Ask yourself what you can do to find fulfillment.

Maybe the giant of self-doubt has you around the neck, and you feel like you're paralyzed with fear. If so, recognize this and confront this giant head-on.

Is your business struggling with the giant of innovation? Take a page out of our DeltaV playbook, and be a constant disrupter.

Are you facing the giant of skeptical management? Set up a series of one-on-one meetings with influential leaders and team members to turn the culture.

What about your competition? Do they seem like a giant too tough to overcome? If so, take some time to develop clear understanding of how you stack up against them. And look for that one area where you might gain a foothold in the industry.

Does success seem elusive and like a giant that taunts you from afar? Keep doing the right thing every day. Move one step closer to your goals, and in time, you'll be amazed at where you end up.

Turning giants is hard. But it is rewarding. Because with each giant you turn, you discover more about yourself and those you lead. You stretch the limits of your human capacity and push others to a higher level. And this makes it worth the effort.

ACKNOWLEDGMENTS

I have been blessed to have so many wonderful people who provided guidance, wisdom, friendship, and love along the way. Let me begin by thanking my parents, whose life was a real example of turning giants. They gave me a value system and a work ethic and, most importantly, their support and love. I also have the world's best sister, Marianne. My wife, Charlotte, has been my unwavering supporter through all the twists and turns of our fifty-plus years of marriage. Our three children, Jim, Dan, and Jenny, are each unique and have provided us with great memories of their childhood and countless moments of pride as adults. We have four grandchildren, Olivia, Jake, Ellie, and Taylor. Retirement has enabled us to enjoy them and spoil them.

I would also like to thank my very first boss, Armando Pasetti. I had a summer and weekend job working for him in a salami factory and retail store. He introduced me to the idea of demanding but fair. My high school years centered around Joe Schulte, who was both a math teacher and a drama director. I loved his classes and acted in several of his plays. At Washington University in St. Louis, Professor

Bill Murphy was someone I admired. Dr. Murphy taught many of my system science classes while consulting with McDonnell on America's space program.

I had three mentors at Monsanto. First was Lee Vehige, who was assigned as my mentor when I first went to work there. He patiently explained everything to me and guided me through the transition from university to the real world. Next is Stan Weiner and his wife, Marilyn. Stan was the engineer everyone went to for technical advice. He had an irreverent sense of humor. Marilyn helped both Charlotte and I get through the life-changing event of having our first child. Another important person was and is Greg McMillan. Greg is a legend in process control, and fortunately, he too went on to help Emerson and still does today. When I left Monsanto, Stan and Greg gave me a retirement party. Their gift was a record album of the cheesiest elevator music. Monsanto subscribed to a service at the time called Muzak, which piped in music to the office. What a productivity boost there was listening to the Hollyridge Strings play "I Can't Get No Satisfaction"!

At Pritchard, Bill Hockersmith was my boss and a great man, literally. He was a huge man. Shaking hands with him was like putting your hand into a catcher's mitt. He taught me a lot. At Beckman, I met Bob Kase. Bob and I were assigned to introduce Beckman's control products into Houston. We have become lifelong friends. The products we sold were not ready for prime time, and we met many times at the end of the day to compare notes and commiserate with each other.

At Rosemount, I had a very important mentor named Gene Schnabel. Gene was head of finance, and I turned to him for help in understanding P&L and balance sheets. He took time to help me learn. His help enabled me to aspire to general management. Of course, Vern Heath was my hero. Vern was a childhood polio victim and turned that giant to go on to found Rosemount and lead it. I learned so much from him about being a leader and the importance of integrity.

At Emerson, I would like to thank Joe Adorjan. Joe was a strategic genius who drove Emerson's move into the process automation business. He also showed confidence in me when he asked me to go to Austin to lead the newly combined system businesses. Chuck Knight, to me, was a model CEO. I would watch how he handled issues and marveled at his ability to find the holes in what you were presenting and ask strategic questions. I always thought that presenting to him was going to the Super Bowl. I have kept and treasured the handwritten notes of congratulations he sent to me.

When Joe Adorjan left Emerson, Bill Davis took his place. Bill was my boss during the initial development time of the DeltaV project. I truly appreciated his support. After Bill came David Farr. David presided over the formative years of DeltaV and PlantWeb. I thank him for his support and have many fond memories of turning some giants together. I also owe a lot to Kathy Button Bell, who first was hired as a marketing and branding consultant and eventually became chief marketing officer of Emerson. Kathy helped us to go boldly into unconventional advertising and marketing. She was also a forceful leader in building a technology brand for Emerson.

I would like to thank two members of my staff for being the ones who would come to me and tell me that I messed up big time and need to do damage control. Dave Hunter not only ran our global sales organization but also contributed so much more. Tracy Thompson was my chief financial officer. Tracy had an incredible way of seeing financial troubles coming before they became truly troublesome. Dave and Tracy are lifelong friends, and I see them both often.

I owe a debt of gratitude to the Forbes team, for guiding me through the giant of writing a book. They are true professionals.

Finally, I would like to thank all of the engineers and technicians who work in automation. They have such a tremendous influence on the performance of the organizations they work for, yet their work largely goes unheralded. They work in a noble profession and deserve recognition.

ABOUT THE AUTHOR

John Berra received a BS in Systems Science and Engineering from Washington University in 1969 and began his career as a control engineer at Monsanto. In 1976, he joined Rosemount, where he held several management positions, including president of the Industrial Division. He was named president of Fisher-Rosemount Systems in 1993 and, in 1999, was promoted to senior vice president and process group business leader for Emerson Electric. In 2008, he was named chairman of Emerson Process Management. He is currently retired.

Mr. Berra has contributed not only to the success of the companies he's worked for but also to the automation industry. He drove the development of the hybrid addressable remote transmitter (HART) protocol. He was also an early proponent of an all-digital fieldbus and served as the chairman of the board of the Fieldbus Foundation from its creation in 1994 to 2010. From 1988 to 1990, he was chairman of the board of the Measurement, Control, & Automation Association.

Mr. Berra served on the board of directors of Ryder System, Inc. (NYSE:R) and National Instruments (NASDAQ:NATI).

Mr. Berra has also received the Emerson Electric Technology Leadership Award, the Washington University Alumni Achievement and Distinguished Alumni Awards, the Lifetime Achievement Award from ISA, and the Frost and Sullivan Lifetime Achievement Award. He was named one of the fifty most influential industry innovators by *InTech* magazine, and he was voted into the Process Automation Hall of Fame.